Basaltic Flows
in
Ghat Section

Basaltic Flows *in* Ghat Section

Dr. Prashant Uttamrao Terker

Oxford Book Company
Jaipur ▪ New Delhi

Contents

Preface

This book titled "Basaltic Flows in Ghat Section" deals with the said subject in its totality. To demarcate the basaltic lava flows and to study those traverses were taken along the ghat road. The contacts between the overlying and underlying flows and their top and bottom are clearly visible along the road cut. The road cuts happen to be fresh and are suitable for photographic display. To determine the order of disposition of flows to top along with their lateral extent, the L sections of the ghat were also prepared. Petrography is a branch of petrology that focuses on detailed descriptions of rocks. Someone who studies petrography is called a petrographer. The mineral content and the textural relationships within the rock are described in detail.

These book is suitable for geophysicists, geologists, scientists, universities, organizations, teachers and students.

Dr. Prashant U. Terker

1

INTRODUCTION

The end of Cretaceous era was marked by catastrophic event which was manifested by volcanic activity which resulted in the form of outpouring of enormous quantity of basaltic lava that covered hundred of sq. km of Western; Central and South India. The term Deccan Trap was first introduced by W. H. Sykes in 1833 after Swedish word Trap/Trappa which means step like or terrace like formation. (Fig.1.1) After Igneous and Metamorphic complex of Archean era in Peninsular India (Fig.1.2) this is the second most extensive geological formation. Deccan Trap encompasses today area of about 516000 square km. covers large part of Kutch, Saurashtra, Gujarat, Maharashtra, Madhya Pradesh and Karnataka. Except East parts of Nagpur district whole of Chandrapur; Gadchiroli and Bhandara districts and southern part of Sindhudurg district, all other districts of Maharashtra are occupied by Deccan

Trap. Thus Maharashtra state can be considered as house of Deccan Trap.

Krishnan (1968-p.406-407) has mentioned perfect horizontality of "Deccan Trap Basalts, but this does not hold true for basalt to basalt contact but this corresponds to contact between Deccan trap and Gondwana rocks. Because of undulating nature of ground over which the flows are overlying the total thickness of Deccan Trap is also variable. The trap has maximum thickness of 3000 meters near Bombay coast. At Girnar in Gujarat the Deccan Trap is estimated to have thickness of 1070 m. The section at Toranmal of west Satpuda region is about 835m thick; at Mahabaleshwar it is 1700 m thick, and at Matheran it has a thickness of 850m. As one moves towards east and south the thickness of Deccan Trap goes on decreasing. At Chikhaldara section of North Vidarbha region the thickness of Deccan Trap is 625 m. The thickness of Deccan Trap in the bore hole drilled near Salbardi Hot spring in Amravati district of Maharashtra has thickness of 604m. At Amarkantak in the east the thickness is 150 m. At south near Belgaon in the state of Karnataka the thickness is 60 m.

The Deccan traps are made of piling of several lava flows one above the other, and each lava flow is ranging from few meters (7m) to as much as 70m. Maximum as has been traced in a drill hole located near Salbardi hot spring (S.K. Sengupta ;S.S. Deshmukh 1996) where 26 lava flows are exposed in a vertical span of 604 m. A bore hole at Bhusawal, revealed 29 lava flows in a vertical span of 370m having average thickness of 12 m. In Chhindwara district of Madhya Pradesh 16 lava flows have been identified with an average thickness of 21 m.in a vertical span of 336 m. where as in the Sausar tehesil of same district in a vertical span of 120 m. 7 lava flows have been revealed with average thickness of 17m. (Krishnan P 406-407) Two different types of lava flows have been recognized they are pahoehoe or ropy lava and the aa or block lava. Pahoehoe lava is mobile lava and solidifies with smooth and glazed surface having rope like wrinkles therefore called as ropy lava. When flows are made of several small units the flows are called as compound (Amygdaloidal) unit. As each unit possesses criteria of demarcation of flows with irregular top surface B.M. Karmarkar et.al. (1994) called such flow units as Thin and Thicker irregular Amygdaloidal Basalt flow. Simple flows are uniform over large area and are made up of single units. Generally pahoehoe characters are exhibited by compound flows and are predominant in Dhule, Jalgaon; Nasik; Pune; Ahamadnagar, Thane and Kulaba districts, where as rest of the region of Maharashtra simple lava flows are predominant.

The compact Basalt flows, without exception, occur as thick extensive flows characterized by regular tabular form, their tops and bottom being plane surfaces. They can be traced for a distance of upto 20 km and larger ones cover a few hundred kilometers of area. However, their thickness varies greatly (B.M. Karmarkar 1974)

23 flows in the Sinhagad Hill 20 km. south of Pune varied in thickness from 4.5 m to 70 m. Compact basalt flows do not show any other form, the extensive lateral spread, large thickness and high l/t ratio of compact (l = lateral Extent; t = thickness of flow) basalt clearly indicate that they are the products of fissure type of volcanicity characterized by floods of enormous quantities of fluid basaltic lavas that spread over a few hundred square kilometers as thick horizontal sheets.

Amygdaloidal basalts on the other hand show great variation in the form and are characterized by much smaller size. More commonly the larger of the amygdaloidal basalt flows are only up to 1km in lateral extent and up to 10 m thickness. (A.V. Tejankar, 2002) Such flows usually do not have regular form as their tops and bottoms are not regular plane surface and their thickness varies because of irregularities in their tops and bottoms.

The amygdaloidal basalts also commonly occur in thick piles of small irregular flows with limited lateral extent. They show pinching and bulging and sudden change in their thickness very often. Traced laterally they are seen to pinch out over small distances. Due to bulbous and lenticular form with irregular top and bottom surfaces sloping in all directions they exhibit irregular form, which results into formation of irregular flow junctions. In many areas of Deccan Trap the intermixing of thin and thick irregular amygdaloidal flows is seen (B.M. Karmarkar 1974, A.V. Tejankar, 2002).

Though the amygdaloidal basalt flows are the products of some fissure type of volcanic activity, the supply of lava was much restricted which is clearly seen from their comparatively limited size and the lavas were more viscous. B.M. Karmarkar (1974) based on the differences in their field characters grouped amygdaloidal basalts into three groups.

A] Tabular flow of amygdaloidal basalts, less thick less expensive and less regular.

B] Thin irregular flows of amygdaloidal basalt.

C] Thick irregular flows of amygdaloidal basalt.

Though it has already been mentioned that volcanic vents indicating central type of volcanicity took place in Deccan Trap (C.V. Agashe 1968; R.B. Gupte 1971). However, no flow have been formed by central type of activity as found anywhere (B.M. Karmarkar 1978). One of the possibilities could be that the lava out poured during the volcanic activity occurred only at higher levels and have now been denuded away. This is supported by the fact that in the Sinhagad, Ajinkyatara Sections the youngest flows have been penetrated by these vents. It therefore appears that central type of activity was prevalent in the closing stages of Deccan Trap volcanicity.

Fig. 1.1: Distant view of Deccan Basalt Flows of Imampur Ghat

Literature Survey:-

The Deccan Trap Province of Western India is one of the largest continental flood basalt provinces of the world. The province covers an area about 516000 sq.km. In Western and Central part of India (Fig. No. 1.2 – A, Fig. No. 1.2 B) and has a thickness of almost 3.5 km in the western region. The succession of flows reaches 2 km. of thickness near Bombay coast and thins out towards east. Due to its huge spread and variety in geological features and structures it has attracted attention of the geologists across the globe. The geological investigations made by experts for various engineering sites in various parts of Deccan Trap regime have revealed that there is considerable variation in its field characters which remained unexpressed by earlier workers. To undertake further studies various Government organizations like ONGC, GSI, GSDA and few Geology departments at university and college level under took further studies of basalt after Indian independence.

The literature survey shows that the earliest work on Deccan trap was done by captain Dangerfield (1824) who prepared geological Map showing outcrops of basalts of Malwa Plateau of Madhya Pradesh. Subsequently a valuable contribution was made to the knowledge of Deccan Trap formation by Captain Thomas John Newbold Assistant Commissioner of Kurnool who surveyed the entire trap along with the distinguished geologists i.e. Voysey; Christie; Benga and Malcolmson and presented his paper to the "Royal Asiatic society" in 1844 on "Summary of the Geology of Southern India." Significant contribution was made by Blandford who studied Deccan Trap formation of Western India (1856); The geology of portion of Kutch (1867); The intertrappean

Beds of Western and Central India (1867); The geology of the Tapti and lower Narbada valleys and some adjoining districts (1869); The geology of Nagpur (1872). P.N. Bose (1884) is the first Indian geologist who presented paper on Geology of the lower Narbada valley between Nimawar and Kawant and he is also credited for bringing a microscope into geological survey and initiating study of rocks under the microscope. L.L. Fermor and C.S. Fox (1916) published paper on 'The Deccan Trap Flows of Linga, Chhindwara District' and were the first to map flow by flow and pick up individual flows over an extended area displaying gentle anticlines and synclines, as well as small faults. L.L Fermor (1920) identified different types of ultrabasic rocks derived from basaltic magma by gravity differentiation. For the first time H.S. Washington (1922) studied the petrochemistry of Deccan Basalt flows and presented the normative values of these flows. Subsequently K.K. Mathur and P.R. Jagapathy Naidu (1932); Ratan N. Sukheswala and Arie Poldervaart (1958); A.C. Chattarjee (1961); published their research papers on Deccan Trap and associated formations. The palaeomangnetic studies of Deccan Trap basalt flows and dykes were carried out for the first time by P.W. Sahastrabudhe (1963), J.E. Beane; C.A. Turner; P.R. Hooper, K.V. Subbarao and J.N. Walsh (1983) published a paper on stratigraphy of western gnat basalts in Deccan flood Basalts, (Mem. Geol. Soc. of India No10 PP.117-134). K.G. Core and C.J. Hawkesworth (1985) published a paper on stratigraphy of the Deccan Traps of Mahabaleshwar; Western Ghats of India in Journal of petrology V 26 (PP.355-377).

The work on structural and tectonic aspects of Deccan Trap was also carried out by J.B. Auden (1949) on the Dykes in Western India and their relationship with Deccan Trap. Similarly George P. Walker (1969) published his research paper on structures observed and their interpretation on Deccan Trap Distribution; (Mem. Geol. Soc. of India No. 43 P.P. 367-395;) C.S. Raja Rao; Y.S. Sahastrabudhe; S.S. Deshmukh (1978) Published their paper on Distribution Structures and Petrography of Deccan Trap India. (Mem. Geol. Soc. of India No.43 p.p. 401-414.) On the basis of their form B.M. Karmarkar (1978) divided the Deccan Trap Basalt flows in five groups. J. Mahoney et. al. (1982) carried out an isotopic investigation of the Deccan Trap flows of Mahabaleshwar to know the origin of flows. K.K. Mathur et.al. (1926) stated for the first time about the Magmatic differentiation of Deccan lava flows in Mount Girnar in the state of Gujrat. P. Krishnamurthy et.al. (1986) carried out mineralogical and chemical studies on alkaline basaltic rocks of Kutch, Gujrat.

Origin at Deccan volcanic province has always been debated. The debate concerns weather the magmas formed by giant mantle plume, normal plate tectonic process or impact of large extra terrestrial bollide.

K.K. Mathur and P.R. Jagapathy Naidu (1932) for the first time published the paper on volcanic activity of coastal tracts of Bombay – Salsette and Bassein in which they have pointed out varied intrusive rock types distinct from Deccan Trap lava with contrast chemical composition in Malviya Commemoration volume (P.P 787-804). It was very difficult to explain the entire phenomenon of volcanicity on earth surface therefore many geologist tried to correlate the volcanicity with the theory of plate tectonics. T.R. McGetchin et.al. (1979) published paper on plateau uplifts made and mechanism where relationship between uplift mechanism and tectonic setting is widely discussed (Mem. Geol. Soc. of India No.43 PP 591 -600). Dan McKenzie (1984) published a paper on 'possible Mechanism for epierogenic uplift in which he stated that one of the major causes for such major upliftment may be by the hot upwelling mantle. (Nature 1984 V 307 PP616-617). However it was

K.C.A. Burke and J.F. Dewey (1973) introduced the plume hypothesis and were first to apply to explain flood basalt mechanism. (Jour. Geology V 81 P.P.406-433) further they have also stated that when a plume rises beneath the continent resulting into upliftment of surface upto two km. in its centre, and fracturing of the dome may result in massive amount of basaltic volcanism.

Plume hypothesis provided a frame work in which many features of geology of non-orogenic continental areas could be considered.

However, there are several authors who have doubted the plume theory as per as origin of Deccan basalts flow is concerned. H.S. Washington (1922) Sir Archibald Geikie (nature 1880); B.M. Karmarkar (1978); Parthsarathy (1978) ; A.V. Tejankar (2002) ; S.D. Doke (2013) and Chandrashekharan noted that Deccan Basalts predominantly erupted through fissures in the crust and proposed a model of shallow mantle melting in which pre-existing rift zones were reactivated and magma simply poured out of fissures.

Considering the research papers of earlier workers, in the present investigation attempt has been made to decipher the nature of volcanicity that has produced basalt flows in time and space. Most of the workers attributed Deccan Trap as the product of fissure type of volcanicity, however, Agashe and Gupte (1968; 1971) have shown that in addition to fissure type, central type of volcanicity also existed in some parts of Deccan Trap area of Western Maharashtra. Subsequently on the basis of systematic studies of the field characters of basalt flows exposed in road cuts and railway cuts taken form ghat sections from M.S.L. 40 m. to

M.S.L. 1220 m over a large area of Deccan Trap formation Gupte et.al. (1971), Karmarkar (1974; 1978) S.R. Kulkarni (1975), Marathe (1976, 1980), P.S. Kulkarni (1984); A.V. Tejankar (2002); S.D. Doke (2013) have shown that there is distinct variation in the nature of volcanicity at different times and at different places in Western Maharashtra, Khandesh and Marathwada region of Maharashtra state. P.K. Thorat (1996); R.K.

Sharma (1996) and Sudha Vaddadi (1996) have reported the occurrence of a definite feeder system of lava flows in some parts of Pune and Ahamadnagar districts of western Maharashtra. Hence the present work is concentrated on these lines for Marathwada region and Western region of Maharashtra state (Fig. 1.3).

The Problem:-

The Geological researchers and scientists have given their attention in the major parts of western Maharashtra and very little focus is given on the field characters of basalt flows exposed in central Maharashtra. Therefore, with view to determine the nature of volcanicity in the western part of Maharashtra adjacent to Aurangabad and Beed Districts in Marathwada region, a detailed study of the basaltic lava flows exposed in four ghat sections was taken from Ahmadnagar District.

For carrying out systematic studies of basaltic lava flows adjacent to Marathwada region Ahamadngar district was found to be suitable area (Fig No.1.4) where four steep ghat sections facing Godavari valley, along the North and North East border of Ahamadnagar district, occurring at higher levels with excellent rock sections are available for observations. Out of four ghat sections three are available in Ahamadnagar district where as Garmatha ghat section is available in the Beed district of Marathwada region. In all ghat sections excellent exposures of basalt flows are available from

M.S.L. 605 m. to M.S.L. 778 m. and these ghat sections are fairly evenly spaced and are found to be quite suitable to get comprehensive idea of the basalts of the area.

Area of Present Investigation:-

In present investigation an attempt has been made to determine nature of volcanicity in central part of Deccan trap formation i.e. In Ahamadnagar (Fig.No.1.2-A) and Beed (Fig.No.1.2-B) Districts of Maharashtra. There are four ghat sections selected from central part of Deccan trap formation. Their location is given in Table No. 1.1.

Table No. 1.1
Location of the Study Area

Name of District	Name of Ghat Section	Base of Ghat Section with MSL & Chainage	Latitude longitude	Top of Ghat Section with MSL & Chainage	Latitude Longitude	Height / Total Vertical Section of Ghat in Meter	Total Length of Ghat Section in Km
	Imampur	Pandhricha Pool - MSL 605 mCh. – 24.300	N 19^{0}16'256" E74^{0}49'929"	Imampur – MSL 723 m Ch. – 20.800	N 19^{0}15'821" E74^{0}49'595"	118 meter	3.5 Km
Ahmadnagar (Chainage from Ahmadnagar Town, onwards)	Vambhori/Dongargan	Dongargan MSL 613 m Ch. – 21.500	N 19^{0}14'527" E74^{0}45'509"	Shendi – MSL 719.50 m Ch. – 19	N 19^{0}11'573" E74^{0}45'740"	106.50 meter	3.5 Km
	Sautada	Jamkhed MSL 627.43 m Ch. – 76	N 18^{0}46'350" E75^{0}20'320"	Boundary in between Beed and Ahmadnagar district – MSL 777.43 m Ch. – 82	N 18^{0}47'06" E75^{0}21'022"	150 meter	7.0 Km
Beed (Chainage fromBeed Town, onwards)	Garmatha	Dongar Kinhi MSL 655 m Ch. – 50	N 18^{0}55'48" E75^{0}26'42"	Bhatepura – MSL 790 m Ch. – 46.50	N 18^{0}54'62" E75^{0}29'215"	135 meter	3.5 Km

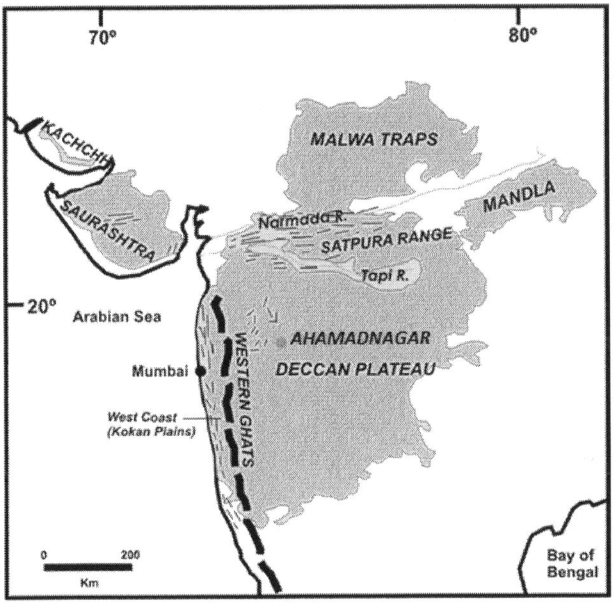

Fig. 1.2 - A : Area Covering Deccan Trap Formation in Ahamdnagar

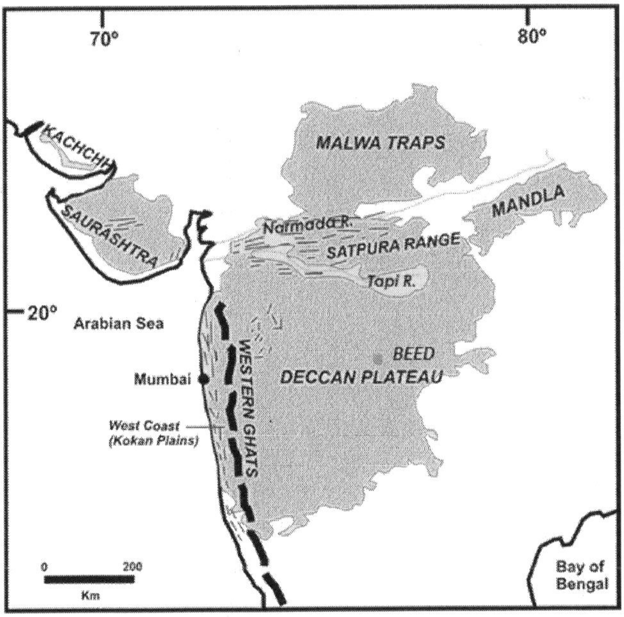

Fig. 1.2 - B : Area Covering Deccan Trap Formation in Beed

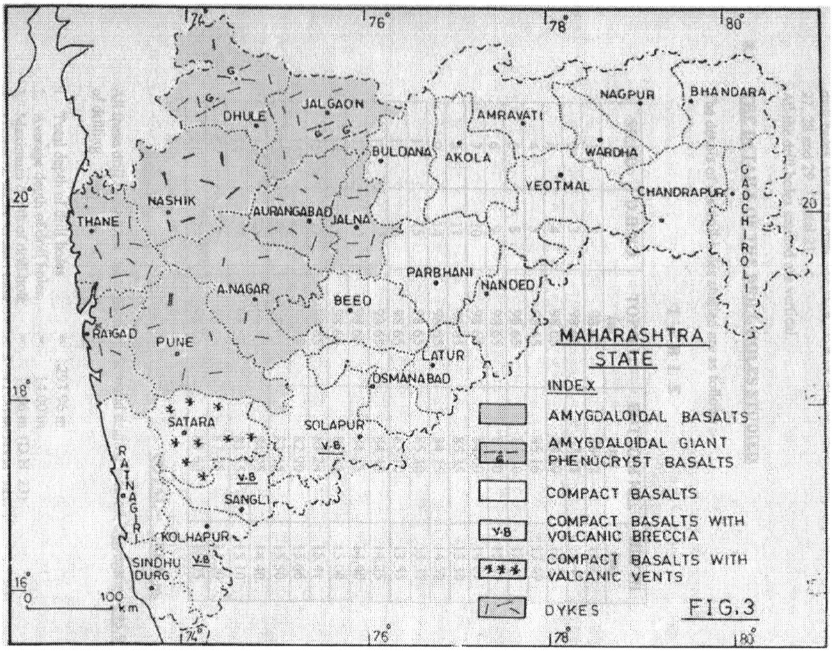

Fig. No. 1.3 Regional Distribution of Basalt Flows

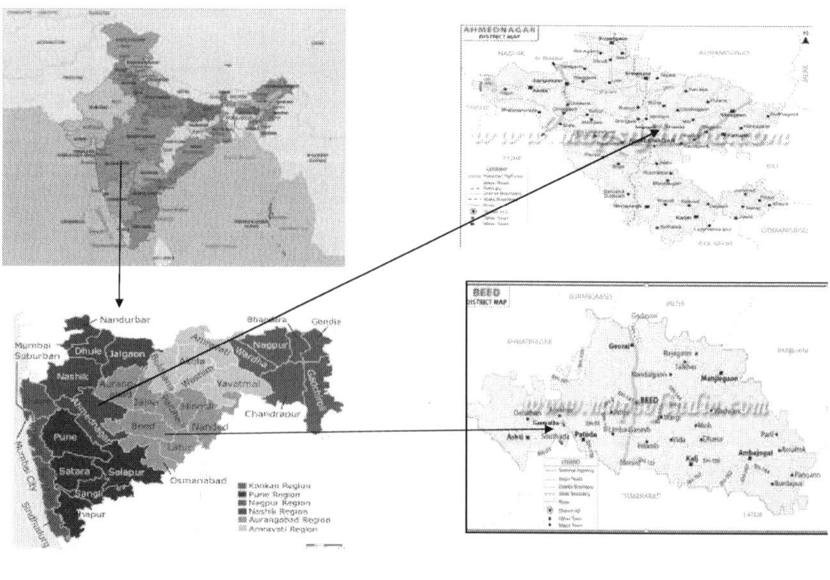

Fig. No. 1.4 : Location Map of Study Area

Imampur ghat section is exposed at a distance of 24.300 km. from the township of Ahamdnagar towards North. The ghat section is exposed on Aurangabad-Ahamadnagar-Pune highway. Vambhori-Dongargan ghat section is exposed at a distance of 21.5 Km. from the township of Ahamadnagar. The ghat section is exposed on Ahamadnagar- Shendi- Vambhori road. It is approachable by all weather roads from Ahamadnagar. Sautada ghat section is exposed at a distance of 76 km. from the township of Ahamadnagar and is very close to Jamkhed Tehsil of Ahamdnagar district. The top of ghat section demarcates the boundary between Ahamadnagar and Beed district of Marathwada region. It is onway from Ahamadnagar to Beed via Jamkhed and is approachable by all weather roads from Ahamadnagar.

Garmatha ghat section is exposed at a distance of 50 km. from the township of Beed and is in Marathwada region, it is exposed on Beed- Rajur-Navgan-Mahendarwadi road and it is approachable by all weather roads from township of Beed.

The Methodology of Investigation:-

A] Study of field characters of Basalt flows:-

As shown by earlier workers (Gupte 1974); Karmarkar (1974); Kulkarni, S.R. (1975); Marathe (1976); Kulkarni P.S. (1984); Tejankar A.V. (2002); Doke S.D. (2013) where more emphasis has been given on the field studies of Deccan Basalt flows because it gives clues to the nature of volcanicity. Therefore much emphasis has been given to the studies of field characters of basalt flows exposed in all four ghat sections of the study area. For this purpose traverses were taken from the foot of the ghat up to its top. The traverses were taken in summer season between 2013 to 2017 when road cuts remain free from cover of vegetation and moss. During the field work individual basalt flows were demarcated and traced in the cutting along the road to determine their extent, thickness and field characters. The vertical variations from top to bottom of each flow were critically studied. Wherever possible the vertical thickness of the basalt flows, their chronological sequence in the vertical stretch of ghat section is depicted.

Numbering of Basalt Flows:-

While numbering the flows, system adopted by Karmarkar (1974) is followed. Due to lateral spread of many flows and gap in between two cuts it as not possible to correlate the flows. Many times a flow is overlaid by two flows, lying side by side, making it difficult to determine which flow is younger and which is older between them. Therefore, in general, flows have been numbered in ascending order. For the sake of convenience of description and correlation of flows, initial letters of the name of the corresponding ghat section is prefixed to the number assigned to the flow occurring in different ghat sections as given below in **Table No.1.2**

Table No. 1.2

Abbriviations Prefixed for the Ghat Sections

Name of Ghat Section	Abbriviation prefixed	Number of flows demarcated
Imampur	I	32
Vambhori-Dongargan	V	12
Sautada-Jamkhed	SW	10
Garmatha	G	06

B] Petrographic studies:-

To determine mineralogical and textural variation in basalt flows a large number of specimen from the basalt flows were collected and their thin sections were examined under petrological microscope. Critical observations of about 200 thin sections were carried out to determine the mineralogical and textural characters of the basalt lava flows. The relative proportion of different minerals by volume in basalt flows was determined by Dalesse- Rosiwal method.

Anorthite content of Plagioclase phenocrysts was determined by Michael Levy method.

C] Chemical studies:-

Suitable specimens from basalt flows have been collected and chemically analyzed. The major element – Oxides content, trace elements of twenty one basalt flows were obtained by utilizing facilities available at Indian institute of Technology – Powai –Mumbai. The results of these analyses are utilized to determine chemical variations in the basalt flows occurring in all four ghat sections of study area. An attempt is made to determine the nature of magma, type of volcanicity in the study area. The results are compared with the chemical characters of basalt flows occurring in these four ghat sections of the study area.

Previous Work:-

In the study area of four ghat sections no work on this problem has been carried out till date. However, work on the similar lines has been carried out in different parts of Deccan plateau. R.K. Sharma and R.K. Thorat have made a significant contribution by recognizing presence of lava propagation system comprising channels, tubes and tunnels in the compound flow sequence of Pune and Ahamadnagar district of Western Maharashtra. (Memoir Gondwna Geol. Society 1996 P.P.449-460). A brief reference of crystal fractionation in compound lava flows of Dongargan ghat of Ahamadnagar district is made by George P.L. Walkar and S.S. Marathe (1969). (Mem. Geo. Soc. India pp. 367-395) in which they have pointed out the occurrence of platy phenocrysts of Plagioclase Feldspar of size greater than 3 cm. Agashe and Gupte (1972) have located nine central vent type features in south and south-East of Nasik which are parallel to West coast. (Mode of eruption of Deccan Trap Basalt volcanology V 35 pp. 591-

601). Sinha and Tiwari (1964) and Subbarao (1972a 1972b) have shown the Pavagad Complex is a product of vent type volcanic eruption.

To decipher nature of volcanicity that has produced flows is the main aspect of the thesis and has been discussed at length. W.D. West (1959) studied the source of Deccan trap lava flows (Journal. Geol. Soc. of India pp. 44-52). L.V. Agashe and R.B. Gupte (1971) published paper on mode of eruption of Deccan Trap Basalt. (Bulletin Volcanologique V 35 pp. 591-601). B.M. Karmarkar (1978) in the 24 km stretch of Borghat section of Central Railway reported a continuous section of 475 m of basalt flow is clearly exposed and based on their lateral extent and thickness. He divided the flows in five groups. S.S. Marathe (1976) reported 27 dykes exposed in the vicinity of 2.5 km. to 3km. from right to left bank of Gangapur Dam of Nasik district. P.S. Kulkarni (1985) studied different ghat sections of central Maharashtra. He correlated various flows exposed in these ghat sections and found that they are different. This means that the flows building up of one ghat did not reach the adjoining ghat. Chande (1986) studied and located 28 dykes in Gangapur. Vaijapur area in Aurangabad district of Marathwada region and stated that out of 28 dykes 21 are of compact basalt nature and 7 are amygdaloidal basalt in nature. Out of 21 dykes 13 showing feeder nature and remaining are hypabyssal injections.

Ashok Kumar Joshi and A. Jeyaram (1999) mapped the Marathwada region and adjoining parts of Deccan Trap with the help of satellite imageries and have shown the relationship between the axes of high and low gravity zone with major lineaments. B.M. Mahapatra and K.K.K. Nair (1996) studied the origin of a red bole from Buldana Ajanta area and stated that they are consolidated air blown pyroclasts where red bole has basic composition and green bole are having acidic composition. B. Bhaskar and N. Maran (1996) have mapped the entire Marathwada region of Deccan basalt and have pointed out that flows of Vaijapur-Ellora-Aurangabad group comprises dominantly compound lava flows with thickness more than 500 m and gradients of 1:500 to S.E whereas flows of Jalna and Mahur comprises alternate sequence of compound and simple lava flows having thickness more than 80 meters and gradient of 1:150 – 1:500 to SE. (Gond. Geol Magzine Vol.2 1996 pp. 535-536). L.L. Vishwakarma (1996) have reported old volcanic fissure of curvilinear shape of 12km length around Ghotan-Khanapur- Ravthala village in Ahamadnager district where the old volcanic fissure filled with pyroclastic material comprising of bombs representing explosive phase of volcanic activity in the area is seen. SudhaVaddadi and R.K. Sharma (1996) have reported occurrence of xenoliths of hybrid nature in Basic dykes in some parts of Pune and Ahamadnagar districts of West Maharashtra. B.M. Karmarkar : S. R. Kulkarni et. al have given an account on the occurrence of large phenocrysts of Plagioclase at various places in Deccan lava flows and have coined the term 'Giant Phenocryst Basalt' lava flows. (Bulletin Volcanologique Tome xxxv- 4.1971 pp. 965-974) S.S. Marathe, S.R. Kulkarni et.al. (1980) have given the variations in the nature of Deccan trap volcanicity of Western Maharashtra. (Memoir Geol. Soc. of India 3 (1980) pp. 143-152).

R.L. Nichols (1936) coined the term flow units to the Deccan lava flows (Jour. Geology 44 pp. 617-630) however G.P.L. Walker (1972) based on the field characters of western ghat lava flows subdivided the lava flows into compound and simple lava flows. (Bulletin Volcanologique 35 pp. 579-590). There has been Geochemical variation reported by scientists in the lava flows of different locations, however Jyotishankar Rayand, Mihir K. Bose (1996) have studied the genesis of Deccan Basalt lava flows and have pointed out the role of crustal components in the magma in the course of its ascent. (Gondwana Geol. Society spl.Vol.2 (1996) PP 293-29). Whereas T. Sano and T. Fuji (1996) based on Ba/Sr ratio gave an account on chemical variation of uncontaminated (Ambenali-like) Basalts of Western Maharashtra and have pointed that it is due to fractional crystallization at low pressure. Sengupta (1996) studied the petrology and geochemistry of Deccan Trap flows intersected in Salbardi drill hole near Salbardi hot spring in Amravati District of Vidarbha region. The lava pile shows similarity with the composition of OIT and MORB basalts and close compositional similarity with Ambenali and Mahabaleshwar formations of western Deccan basalts. S.M. Godbole and A.K. Chatterjee (1996) studied the petrology and Geochemistry of Deccan Trap flows intersected in the vicinity of thermal spring near Nazardeo in Jalgaon district. The drill hole intersected 465.50 m thick basaltic lava pile comprising 23 flows. The flows are fine grained light green to dark grey in colour showing varying degrees of porphyricity ranging from non porphyritic to high porphyritic with interflow contacts marked by red, green of grey bole beds or highly vesicular tops. The thickness of the lava flow ranges from 2.6 m. to 75.5 m. They have shown that the lava flows are quartz tholeiites with differentiation indices ranging from 24 to 34 indicating that they belong to middle stage of evolution with Mg number ranging from 48 to 58 with significant plagioclase fractionation.

2

GHAT TRAVERSES

To demarcate the basaltic lava flows and to study those traverses were taken along the ghat road. The contacts between the overlying and underlying flows and their top and bottom are clearly visible along the road cut. The road cuts happen to be fresh and are suitable for photographic display. To determine the order of disposition of flows to top along with their lateral extent, the L sections of the ghat were also prepared.

While demarcating the basalt flows along the road cuts the concept of basalt flows is used as defined by Karmarkar [1974] he defined lava flows as 'a mass of lava which irrespective of its size and shape as seen in the field is supposed to have been outpoured in a single uninterrupted eruption which have occurred after the solidification of at least the upper portions of the previous eruptions and not before that portion. This can be clearly demarcated in the field from overlying and underlying masses of lava'.

Criteria used to demarcate flows

1) **Petrological difference:-**

Petrological difference between top portion of underlying flow and bottom portion of overlying flow. The two basalt flows can be easily demarcated in the road cut where outcrops are easily seen and have clear and clean exposure without any cover of vegetation.

2) **Occurrence of red tachylytic basalt along flow junction:-**

In most cases red and rarely green or black coloured tachylytic basalt occur between the bottoms of upper and the top of lower flows or they may occur in the form of intrusions. With the help of these bands one flow can be demarcated from the other flow.

3) **Difference in the nature of outcrop:-**

To demarcate the one flow from other it was very difficult because in some stretches of the ghat section the vertical cuts have not been taken. However, with the help of difference in appearance, in the nature of outcrop of the top portion of underlying flow and bottom portion of overlying flow basalt flows can be demarcated from each other. The top portion of compact basalt flow is being vesicular amygdaloidal and unjointed has always smooth outcrop with some convexity in the slope, whereas the bottom portion of the flow is jointed has always got very rough appearance with almost vertical outcrop. With the help of this difference in the appearance of outcrop two flows can be demarcated from each other. In such cases the flow junction excavations on hill slopes were taken wherever necessary.

4) **Hydrothermally altered vesicular and amygdaloidal top portion:-**

The top portion of the compact basalt flow is always vesicular amygdaloidal and rather greenish and purple in colour due to hydrothermal alteration. Injection of red or green tachylytic basalt also generally occur in the top portion whereas the bottom portion of every compact basalt flow is black in colour and is devoid of vesicles and amygdales and is always dissected by joints. Due to distinct difference in appearance top portion of underlying basalt flow and the bottom portion of the overlying flow, No compact basalt flows in contact with each other, in vertical cut, can be easily demarcated from a distance.

Criteria used to demarcate dykes:-

Due to its particular nature of jointing it is very easy to demarcate dyke in field. During the field work in the ghat sections at Sautada one dyke is exposed and it cut across the jointed compact basalt flows and can be easily demarcated with the help of following criteria.

Though the country rock and dykes are of compact basalt and are jointed there is definite uniformity in the pattern of jointing in the dyke in contrast to jointing occurring in the country rock. The dyke has got mutually perpendicular three sets of joints. Out of these joints two sets are vertical and one is horizontal. Due to these three mutually perpendicular sets of joints, the joint blocks occurring in the dyke always acquire prismatic form. It was also observed that along the margin of dyke the joints are very much closely spaced and the dyke rock is broken into small chips. This helps in demarcating the dyke from the adjacent country rock which do not have definite jointing pattern.

Traverses Taken:-

As the study area comprises of four ghat sections and each ghat section is comprised of different no. of lava flows, therefore, the outcrops and field characters of these flows are described serially as we climb the ghat along the road. The chainages are measured with reference to the chainage 'km' stone located at the respective places of the corresponding ghat section on the road.

Traverses Taken of Imampur Ghat:-

2.1 Geology of Imampur Ghat Section:-

Imampur ghat section is located in the district of Ahmadagar in the state of Maharashtra. It is available on Aurangabad-Ahmadnagar highway. The ghat section is 87.700 k.m away from Aurangabad city towards Ahmadnagar and 24.300 k.m away from Ahmadnagar city while going towards Aurangabad. The ghat section starts appearing from village Pandhricha pool which is situated at the base at the height M.S.L. 605 m The ghat section ends at village Imampur which in situated at top M.S.L. 723 m.

From the low lying area of Imampur Ghat to the top of hills the traverses were taken from vertical road cut of the hills along the ghat. Fresh rock exposures are available for observation at road cuts of ghat.

The traverses were taken from the village 'PANDHRI' which is known as 'PANDHRICHA POOL' at M.S.L. of 605 meters. Latitude N 19^0 16' 256" and Longitude E 74^0 49' 929".

As the lava flows are exposed adjecent to each other (side by side) in this ghat section therefore the actual thickness of the flows varies with MSL.

FLOW NO.: 01

Flow No. 01 is exposed in the dugwell situated at the left side on Aurangabad-Ahamadnagar road when one is facing towards Ahamed- nagar, near the village 'PANDHRICHAPOOL', (MSL 605 m) The flow is of compact porphyritic basalt. It has small to medium sized grains of plagioclase. These phenocrysts appear turbid in colour to naked eyes. Middle portion of the flow exhibits broadly spaced jointing pattern in the well section. The top portion of the flow is of Amygdaloidal basalt and is hydrothermally altered. The lateral extent of the flow is 300 meters and the thickness is of 40 meters. (Fig. No. 2.1)

This flow is exposed at the right side of Aurangabad-Ahamadnagar road highway when one is facing towards Ahamadnagar. (MSL 643 m) This is thin band of Red and Green tachylyte present on top of Flow No.1 [Fig.No.2.1] The thickness of the band is of approximately 0.50 meter only. Green tachylyte is present and is covered by red tachylyte at bottom and top too. The thickness of green tachylyte is ranging from 12 cm to 15 cm.

FLOW NO.: 03

Top of Flow No.2 is overloaded by Flow No. 3 [Fig. No. 2.1] and it is flow of porphyritic basalt. (MSL 643.50 m) The flow has lateral extent of 300 meters and is intensely weathered at some places. It also shows greenish colour at few places due to chlorpheite content. It is also hydrothermally altered due to which it exhibits brown colour. The flow has lenticular shape. At center several horizontal quartz veins are present and they are running parallel to each other. At many places on its surface many vesicles are seen. The lateral extent is of 300 meter and has thickness of 8 meters.

FLOW NO.: 04

This flow is of compact porphyritic basalt. It has well developed lath shaped plagioclase phenocryst. [Fig. No.2.2] The phenocrysts are turbid in colour and are visible to naked eyes. There are few calcite veins present in the flow and the thickness of the veins is of 5 cm and well developed rhombohedral calcite crystals are seen. [Fig.No.2.3] The thickness of the flow is of 6 meters. (M.S.L. 649.5 m) and the lateral extent is of 80 meters. There is no consistency in its joining pattern.

This is thin flow of Amygdaloidal basalt. [Fig. No. 2.4] The flow has thickness of 1.5 meters and is highly weathered. The flow is well sandwiched in between Flow No. 4 and Flow No. 6. There is no consistency in the jointing pattern. The lateral extent of the flow is 90 meters. At some places band of pipe amygdales of 2.5 cm length are also seen. (M.S.L. 651 m)

FLOW NO.: 06

This flow starts appearing from chainage 88.380 km. The flow is compact porphyritic basalt. The flow has thickness of 10 meters. [Fig.No. 2.4] The lath shaped plagioclase phenocrysts are clearly visible to naked eyes. It has lateral extent of 120 meters. i.e. From chainage 88.380 km to 88.500 km. Many horizontal quartz veins running nearly parallel to each other are seen. (M.S.L. 665 m)

FLOW NO.: 07

This flow is compact porphyritic basalt flow. Lath shaped plagioclase phenocryst having length of 1 to 2 cm are visible to naked eyes. [Fig. No. 2.5] Many horizontal quartz veins running parallel to each other are clearly seen. Due to hydrothermal alteration the upper surface appears greenish in colour. It has lateral extent of 56 meters and thickness of 5.5 meters. (MSL 670)

FLOW NO.: 08

The top of Flow No.7 is overlaid by Flow No. 8 and it is Amygdaloidal Basalt flow. It is thin and has thickness of 0.5meters only. [Fig.No.2.5] The flow is weathered at places pipe amygdales of 1 inch length are also seen. Amygdales are filled with secondary minerals like zeolite and calcite. (MSL 670.5 m)

Flow No.9 is a compact porphyritic basalt flow. [Fig.No.2.5] Small sized plagioclase phenocryst are visible to naked eyes and at same places kankar deposition is also seen. The flow is lenticular in shape and has lateral extent of 17 meters. (MSL 672 m)

FLOW NO.: 10

This is amygdaloidal basalt flow starts appearing on top of Flow No. 9 the flow has thickness of 0.5 meter.

[Fig. No. 2.5] Amygdales are filled with secondary minerals like zeolite and calcite are seen. The surface has rough appearance. (MSL 672 m)

FLOW NO.: 11

This flow is compact porphyritic basalt flow. Lath shaped plagioclase phenocrysts are clearly visible to naked eyes. [Fig.No.2.6] The thickness of the flow is of 4.5 meters. The flow is lenticular in its shape and has lateral extent of 56 meters.

This is to be noted that from Flow No. 7 to Flow No.11 all flow are present exactly one above the other and the lateral extent is 56 meters only Flow No. 11 is covered by subsoil surface. (MSL 676 m)

FLOW NO.: 12

The flow is compact porphyritic basalt flow lath shaped plagioclase phenocrysts of 1 cm are clearly visible to naked eyes. [Fig. No. 2.7] The thickness of the flow is 14 meters and the lateral extent is 100 meters is also. At its center an intrusion is seen. (MSL 680)

FLOW NO.: 13

This is Amygdaloidal basalt flow present on the top of Flow No. 12. It has lateral extent of 30 meters, and thickness of 3 meters. [Fig. No. 2.7] The amygdales are filled with chlorpheite and are visible to naked eyes. (MSL 681 m.).

This flow is compact porphyritic basalt flow. The flow has its thickness of 30 meters and lateral extent is of 32 meters. [Fig. No. 2.7] Intrusions of quartz veins are seen into it. (MSL 682)

FLOW NO.: 15

This is irregular Amygdaloidal basalt flow. The thickness of the flow is 3 meters and lateral extent of 32 meters. [Fig. No. 2.7] Its upper portion is covered with subsoil. (MSL 684)

FLOW NO.: 16

The flow is a compact porphyritic basalt flow. The lath shaped plagioclase of 1 to 2 inches is visible to naked eyes. [Fig. No. 2.8] The flow has lateral extent of 3 meters only and thickness of 2 meters only. The flow has convex shape. (MSL 685 m)

FLOW NO.: 17

The flow is present exactly above Flow No 16. The thickness of this flow is 1.5 meters. Its lateral extent is of 6 meters and is convex shaped upwards. From Its bottom (i.e. from top portion of Flow No. 16) intrusion of lava is clearly visible and it has covered top portion of Flow No. 16. This intrusion has its height of 1.5 meter and breadth of 12.5 cm at center and has occupied open space and voids present at the top portion of lava Flow No. 16 and has come downwards in the shape tongue or lobe. [Fig. No. 2.9] Due to outpouring of lava the flow exhibits amygdaloidal structure through its contact margins as well as at center. It exhibits green colour out crop in field. (MSL 686.5 m)

It is present exactly above Flow No. 17. The flow is compact porphyritic basalt flow. [Fig. No. 2.8:2.9] Lath shaped plagioclase phenocrysts are visible to naked eyes. The flow is 1 meter thick and lateral extent is of 6 meters only. Jointing pattern is not seen. (MSL 688 m)

FLOW NO.: 19

The flow is present on flow No 18. The flow is compact porphyritic basalt. The thickness of flow is 3 meters. [Fig. No. 2.10] The lateral extent is of 54 meters. Along the road side it is highly weathered. (MSL 688 m)

FLOW NO.: 20

The flow is compact porphyritic basalt flow. Lath shaped plagioclase phenocrysts of 1 inch size are visible to naked eyes. [Fig. No. 2.11] The thickness of the flow is of 3.5 meters and the lateral extent is of 50 meters at some places the flow is weathered and spheroidal weathering is also seen. (MSL 690)

FLOW NO.: 21

The flow is compact porphyritic basalt flow. Well-developed lath shaped plagioclase crystals are clearly visible to naked eyes. The flow starts exactly above the half way of Flow No. 20. [Fig. No. 2.12] The lateral extent of the flow is of 114 meters and the thickness is 3 to 4 meters. (MSL 691 m)

The flow is red tachylyte (red Bole). The flow starts appearing from chainage 89.790 km. and has lateral extent of 160 meters. It demarcates the contact of Flow No. 21 and Flow No 23. [Fig. No. 2.12] The thickness of the flow is of 0.50 meter. (MSL 692 m)

FLOW NO.: 23

This flow starts appearing at chainage 89.950 km at MSL 684 meters. The flow is compact porphyritic basalt flow. [Fig. No. 2.12] Well-developed lath shaped plagioclase phenocrysts of 6 cm length are clearly visible to naked eyes. (MSL 692.5 m)

FLOW NO.: 24

This is Amygdaloidal basalt flow present above Flow No 23. The thickness of the flow is 1.5 meters and lateral extent is 8 meters. [Fig. No. 2.13] At the contact zone of this flow with Flow No. 23 at few places Amygdales of 3 cm length are clearly seen. These amygdales are filled with chlorpheite. (MSL 693 m)

FLOW NO.: 25

The flow is compact aphanitic basalt, due to intense weathering its upper surface shows pitted appearance. Thickness of the flow is of 2 meters only and lateral extent is of 50 meters. [Fig. No. 2.13] Spheroidal weathering in the flow is also clearly seen at many places. (MSL 694 m)

This flow is present on Flow No 25. This is an intrusion of Amygdaloidal Basalt flow. The intrusion is having tongue shape appearance. The intrusion is from lava Flow No. 25 at its base. [Fig. No. 2.14] The intrusion has width of 30 cm. and height of 1.5 meters. At its center and contact margins it exhibits brown colour. Well-developed pipe amygdales filled with chlorpheitic content with size upto 7 cm. are clearly seen lateral extent of the intrusion is of 2.5 meters. (MSL 696 m)

FLOW NO.: 27

This is flow of Amygdaloidal basalt and has an intrusion of Amygdaloidal basalt flow. The intrusion has intruded from its base and has height of 1.5 meter. [Fig. No. 2.15] Its width is of 15 cm at center. The intrusion at its top has lateral extent of 2.5 meters on both the sides (lateral) and exhibits umbrella like appearance. (MSL 697.5 m)

FLOW NO.: 28

This flow is of compact aphanitic basalt. It is present on Flow No. 27; this flow has thickness of 2.5 meters. [Fig. No. 2.16] Due to weathering at some places the flow appears to be green in colour. Lateral extent of the flow is of 20 meters. At some places it shows spheroidal weathering too. (MSL 699 m)

FLOW NO.: 29

This is a thin band of red tachylyte. [Fig. No. 2.16]

The thickness is of 0.50 meter. It has lateral extent is of 130 meters. (MSL 701.5 m)

The flow is present on Flow No 29. This is compact aphanitic basalt flow. [Fig. No. 2.17] It is lenticular in shape at some places. It starts appearing from ch. 90.100 km. The lateral extent of the flow is of 150 meters. The thickness of the flow is 12 meters. At its top vesicles are also seen and some vesicles are filled with secondary minerals too. (MSL 702 m)

FLOW NO.: 31

This is the last i.e. Fourth flow of red tachylyte exposed in Imampur ghat section. [Fig. No. 2.17] The thickness of the flows is of 0.50 meter. The lateral extent of the flow is of 150 meters. (MSL 718.5 m)

FLOW NO.: 32

This flow is present at the topmost portion of the ghat section this is compact aphanitic basalt flow. [Fig. No. 2.17] The thickness of the flow is 4 meters. Well-developed vertical joints are clearly visible. It is situated at the height of 694 meters and its lateral extent is 80 meters. (MSL 719 m)

Lava Flows of Imampur Ghat Section

Fig.No. 2.1 Flow No.1, 2 and 3.

Fig.No. 2.2 Flow No. 4 Compact Porphyritic Basalt Flow.

Lava Flows of Imampur Ghat Section

Fig. No. 2.3 Calcite Veins In Flow No. 4

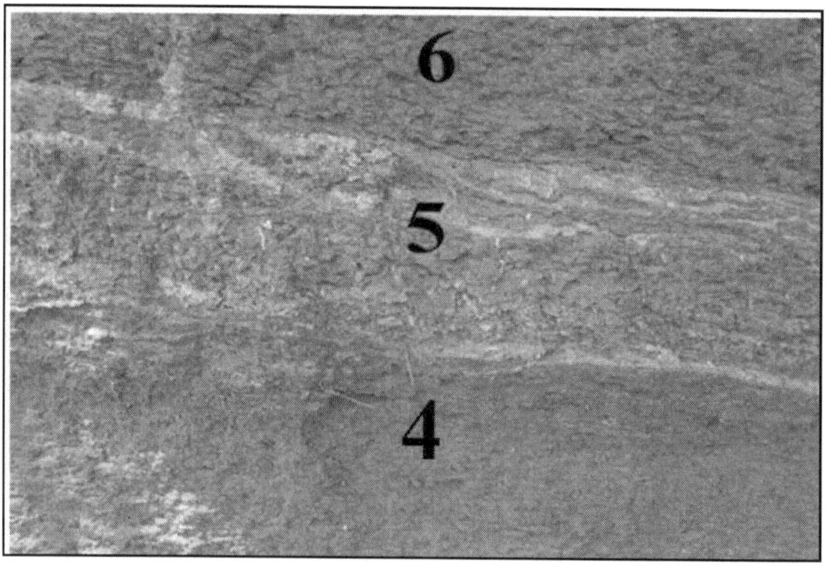

Fig No. 2.4 Contact between Flow No. 4, 5 and Flow No. 6

Lava Flows of Imampur Ghat Section

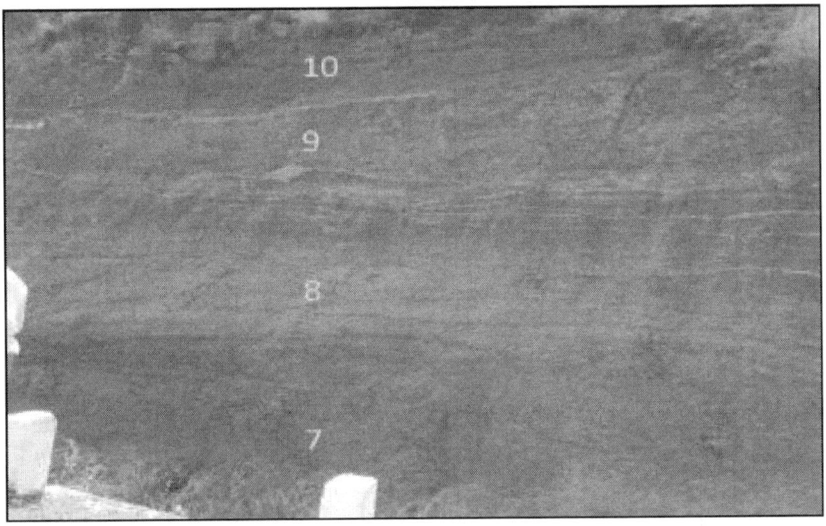

Fig. No. 2.5 Flow No.7, 8, 9 and Flow No.10

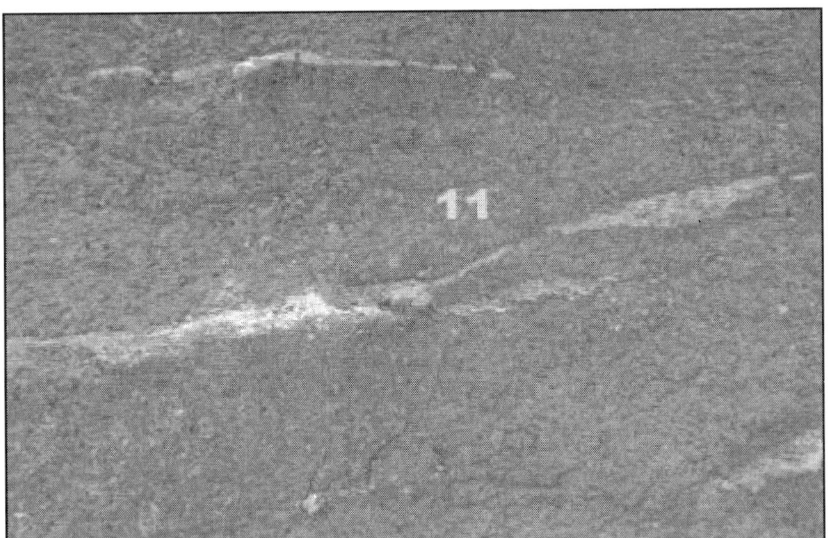

Fig. No. 2.6 Flow No.11 Having Lenticular Shape

Lava Flows of Imampur Ghat Section

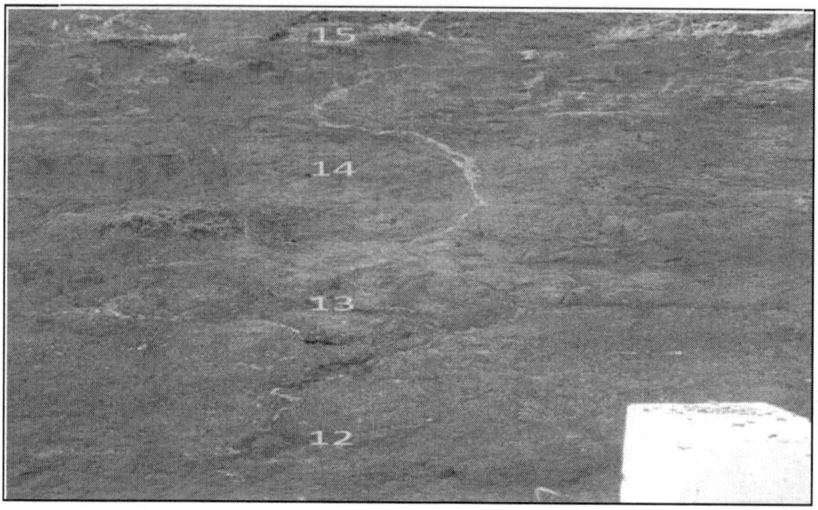

Fig. No. 2.7 Succession of Flow No.12, 13, 14, and Flow No.15

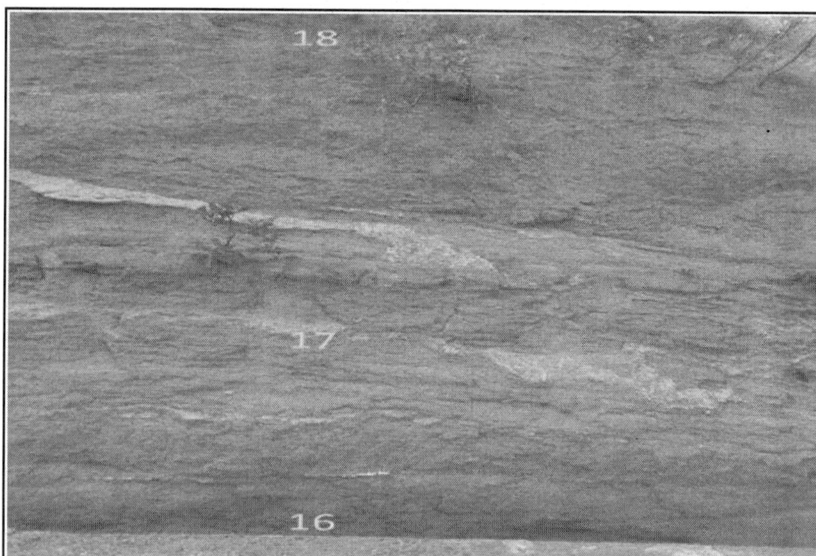

Fig. No. 2.8 Flow No. 16, 17 and 18

Lava Flows of Imampur Ghat Section

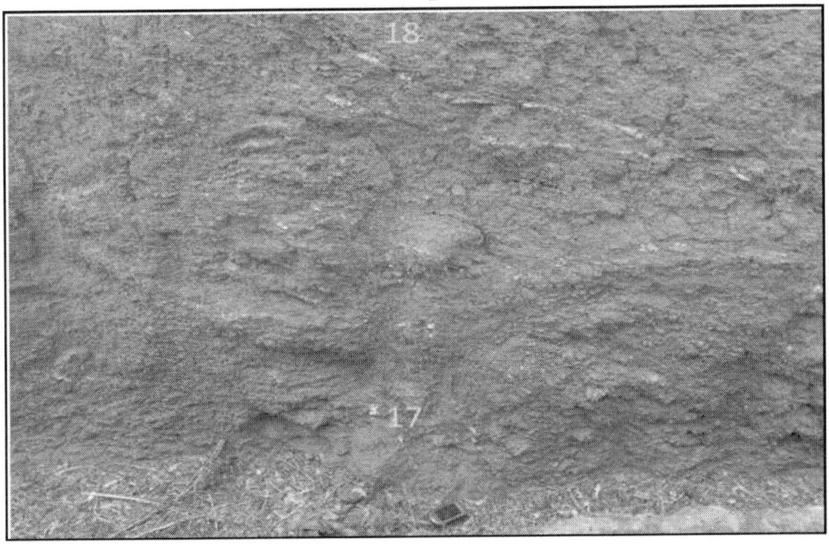

Fig. No.2.9 Intrusion of Lava Flow in Flow No.17

Fig. No. 2.10 Flow No. 19 Porphyritic Basalt Flow

Lava Flows of Imampur Ghat Section

Fig. No.2.11 Flow No.20 Compact Basalt Flow

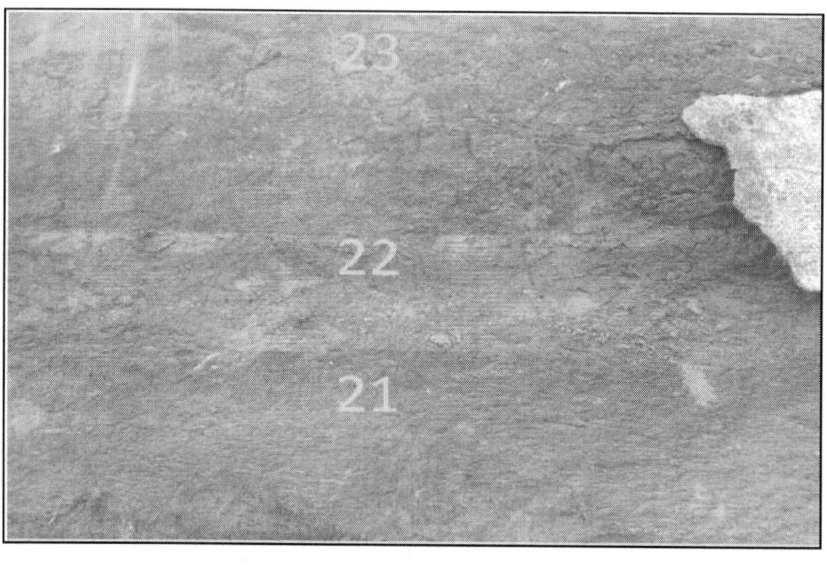

Fig. No. 2.12 Band of Red Tachylyte in between Flow No. 21 &23

Lava Flows of Imampur Ghat Section

Fig. No. 2.13 Flow No. 25 Showing Pitted Appearance Due To Spheroidal weathering

Fig. No. 2.14 An Intrusion of Lava Flow In Flow No. 26

Lava Flows of Imampur Ghat Section

Fig.No.2.15 An Intrusion Having Umbrella Shape in Flow No. 27

Fig. No. 2.16 Red Tachylyte on Flow No. 28

Lava Flows of Imampur Ghat Section

Fig. No 2.17 Band of Red Tachylyte in between Flow No. 30 and 32

Fig. No. 2.18 Field Work in Imampur Ghat

Traverses Taken of Vambhori / Dongargan Ghat

2.2 Geology of Vambhori/ Dongargan Ghat Section

Wamhori Dongargan ghat section is exposed on Ahmadnagar- Shendi – Dongargan - Vambhori road. It is 21.5 km away from Ahmadnagar. The ghat section starts from Vambhori village river bridge which is at base [ch.21.500] M.S.L. 613 m. to Dongargan village at top M.S.L. 719.50 m of [ch.19]

The entire Wambhori/Dongargan ghat section comprises of twelve flows out of which three flows are of closely to broadly space jointed aphanitic compact basalt. Eight flows are thick and are of closely to broadly space jointed porphyritic compact basalt where as one flow is of thin irregular sheet jointed amygdaloidal basalt.

FLOW NO.: 01

The flow starts appearing along the road from ch.21/ 500 m. [M.S.L. 613.10] in the road cut located below the wambhori river bridge when one is facing towards Shendi ie. towards Ahamadnagar [Fig. No.2.19]. The flow is compact porphyritic basalt flow with medium sized phenocryst of plagioclase. The plagioclase phenocrysts are turbid white in colour. The flow is well exposed in the bed rock of river.

As we ascend the ghat section, gradually compact basalt changes into amygdaloidal basalt which indicates the top portion of Flow No.1. The amygdaloidal top portion of the flow is greenish and the amygdales are filled with secondary minerals like silica, calcite, zeolite, chlorphaeite etc. Due to weathering sheet jointing is also developed in the top portion of flow. The exposed thickness of the flow is 27.13 meter.

This flow is exposed on the left side in the road cut between ch.21 km to ch.20.800 km. [M.S.L. 640.243 m to M.S.L. 643.597 m] Thus the thickness of the flow is of 3.35m. The flow is compact porphyritic basalt with medium to big sized phenocrysts of plagioclase. [Fig.No.2.20] The phenocrysts show vitreous luster and have length up to 7 to 8 cm. The flow is closely jointed and top portion of the flow is hydrothermally altered amygdaloidal basalt. The middle portion of the flow shows spheroidal weathering.

FLOW NO.: 03

This is compact porphyritic basalt flow and is exposed in the road cuts between ch.20.800 km. to ch.20.500 km. The exposure is along the left side when one is facing towards shendi/Ahamadnagar. The flow is exhibiting small to medium sized plagioclase phenocrysts. [Fig.No.2.21] Due to hydrothermal alteration the exposed flow show greenish appearance and the middle and bottom portion has suffered spheroidal weathering. The top portion shows closely jointing pattern and the flow is hydrothermally altered to Amygdaloidal basalt. The thickness of the flow is of

8.842 meter. i.e. From M.S.L 643.597 m. to M.S.L 652.439 meter.

FLOW NO.: 04

This flow is exposed at the right side of the ghat when one is facing towards Ahamadnagar. The flow is exposed from ch.20.500 km to 20 km stone i.e. From M.S.L 652.43 m to M.S.L 659.75 meter has thickness of 7.32 meters. The flow is closely jointed and its top portion showing hydrothermal alteration to Amygdaloidal basalt. The Middle and bottom portion show spheroidal weathering pattern. [Fig. No.2.22] White coloured plagioclase phenocrysts are commonly observed in the flow. Sheet jointing pattern is seen in the top and middle portion of the flow.

The flow starts appearing from ch.20km stone to ch.19.950km. i.e. From M.S.L. 659.75m to M.S.L 668 meters. The thickness is of 8.25 meters. The top portion is hydrothermally altered amygdaloidal basalt whereas middle portion is suffered from spheroidal weathering. It is compact prophyritic basalt with small to medium sized plagioclase phenocrysts which are white in colour. [Fig. No.2.23]

FLOW NO.: 06

This flow is of compact porphyritic basalt. The white coloured plagioclase phenocrysts embedded in glassy matrix are clearly visible and show vitreous luster. [Fig. No.2.24] In the middle and bottom portion spheroidal weathering is clearly seen. The top portion of the flow is hydrothermally altered basalt having thickness of 1.82 meters the flow starts from chainage ch.19.950 km to 19.750 km. The flow starts from

M.S.L 667.98 m to M.S.L 685.36 m. Thus the thickness is upto 17.38m. Its flat top is clearly seen at the right side of the ghat section when one is facing towards Ahamadnagar/Shendi.

FLOW NO.: 07

This flow starts appearing at ch.19.750 km at M.S.L .685.36 m and has height of 3.28 m. This flow is also of compact porphyritic basalt with small to medium sized plagioclase phenocrysts. The bottom middle portion of the flow is closely jointed whereas top portion is hydrothermally altered basalt and at some places pipe amygdales are also seen having length 4 to5 cm. The flow also exihibit spheroidal weathering pattern. [Fig. No.2.25]

This flow starts appearing from M.S.L.689 m. This flow is of Amygdaloidal basalt with small to medium sized amygdales filled with silica and zeolites. [Fig. No. 2.26] The flow is highly weathered and has developed closely spaced sheet jointing pattern. The thickness of the flow is 2.13m.

FLOW NO.: 09

This flow starts appearing from M.S.L.691.46m. The top portion of the flow is hydrothermally altered amygdaloidal basalt. [Fig. No.2.26] Spheroidal weathering is commonly seen in the middle and bottom portion of the flow with closely spaced jointing pattern. The flow is compact porphyritic basalt with plagioclase phenocrysts having length up to 6cm. are seen. The phenocrysts of plagioclase exhibit vitreous luster. The thickness of the flow is of 2.75 meters.

FLOW NO.: 10

This flow is of compact aphinitic basalt which starts from M.S.L. 694.20m. The bottom portion of the flow is having closely spaced jointing pattern whereas the top portion of the flow has developed broadly spaced jointing pattern. [Fig. No.2.27] The flow ends at M.S.L. 701.20 m. Thus the thickness of the flow is 7 m.

The flow starts at M.S.L. 701.20 m. The flow is of compact aphanitic basalt. The flow is black in colour. [Fig. No.2.28] Bottom portion of the flow is showing broad space jointing pattern. The top portion of the flow is weathered and is hydrothermally altered. The flow ends at M.S.L. 709.50 m.

FLOW NO.: 12

This flow starts appearing from M.S.L. 709.50 m. The entire flow exhibits black colour. The bottom portion of the flow exhibits three sets of broad spaced joints. [Fig.No.2.29] The lath shaped plagioclase phencocrysts are visible to naked eyes having length up to 8cm. The top portion of the flow is highly weathered. The flow ends at ch.19 km M.S.L.719.50 m.

Lava Flows of Vambhori / Dongargan Ghat Section

Fig. No. 2.19 Flow No.1 Porphyritic Basalt Flow Exposed in Vambhori River

Fig. No. 2.20 Flow No. 2 Porphyritic Basalt Flow

Lava Flows of Vambhori / Dongargan Ghat Section

Fig. No. 2.21 Flow No. 3 Compact Basalt Flow

Fig. No. 2.22 Spheroidal Weathering in Flow No. 4

Lava Flows of Vambhori / Dongargan Ghat Section

Fig. No. 2.23 Flow No. 5 Porphyritic Basalt Flow

Fig. No. 2.24 Hydrothermal Alteration of Top of Flow No. 6

Lava Flows of Vambhori / Dongargan Ghat Section

Fig. No. 2.25 Flow No. 7 Compact Basalt Flow Having Weathered Top Portion.

Fig. No. 2.26 Contact between Flow No. 8 and Flow No. 9

Lava Flows of Vambhori / Dongargan Ghat Section

Fig. No. 2.27 Development of Closely Spaced Jointing Pattern in Flow No. 10

Fig. No. 2.28 Development of Broadly Spaced Jointing Pattern in Flow No. 11

Lava Flows of Vambhori / Dongargan Ghat Section

Fig. No. 2.29 Weathered Top Portion of Flow No. 12

Fig. No. 2.30 Field Work In Vambhori / Dongargan Ghat

Traverses Taken of Sautada Ghat

2.3 Geology of Sautada Ghat Section

Sautada ghat section is exposed on Ahmadnagar-Jamkhed- Beed

The ghat section is available in outskirts of Jamkhed at a distance of 76 km. from the township of Ahmadnagar. The ghat section starts appearing at M.S.L. 627.43 m. at its base and its top demarcates the boundary between Ahmadnagar district and Beed district of Marathwada region M.S.L. 777.43 m.

The traverses were taken from the base i.e. from the outskirts of Jamkhed town to the top of the hills and from the vertical road cut of the hills along the ghat section. Fresh rock exposures are available for observation at road cuts of the ghat. The top of the ghat section demarcates the boundary between Ahamadnagar district and Beed district of Marathwada region.

FLOW NO.: 01

This flow is well exposed in the dugwell and is a thin band of green tachylyte. [Fig. No. 2.31] The thickness of the flow is 0.57 m only. (M.S.L.627.43 m).

The flow starts appearing in the dugwell above the thin band of green teachylytic basalt. However it is clearly visible along the road (M.S.L. 628.00 m) and it appears at right side of the road when one is facing towards the top of the ghat section i.e. towards Sautada.

The flow section appears fresh in the dugwell. The flow is of Amygdaloidal basalt and is green in colour, due to hydrothermal alteration. [Fig. No.2.32] The amygdales are filled with secondary minerals like quartz, zeolite and calcite. Due to weathering the top portion of the flow has developed sheet jointing pattern.

FLOW NO.: 03

The flow starts appearing (M.S.L. 628 m) on the road cut at the right side of the ghat section when one is facing towards the top of ghat section Sautada. This flow is compact aphanitic basalt. [Fig. No.2.32] The closely spaced jointing pattern is clearly seen in the top portion of the flow. The thickness of the flow is 9.19 meters.

FLOW NO.: 04

The flow is well exposed at Moharedwadi chowk. The flow is Amygdaloidal Basalt flow. The top of the flow has developed sheet jointing pattern. The small to medium sized amygdales are filled with secondary minerals like silica; calcite and zeolites. [Fig. No. 2.33] The closely spaced jointing pattern is also seen. The thickness of the flow is 8.229 meters. (M.S.L. 645.42 m)

The flow starts appearing at the left side as we proceed in ascending pattern in the ghat section. This flow lies on the top of Flow No.4 and the flow junction is clearly seen. [Fig. No.2.33] This flow is compact aphanitic basalt flow. The top of the flow is hydrothermally altered amygdaloidal basalt having closely spaced jointing pattern, whereas bottom portion has developed broadly spaced jointing pattern. Spheroidal weathering is commonly visible in the middle portion of the flow. The thickness of flow is of 50 m. (M.S.L. 695.12 m)

FLOW NO.: 06

It is reddish brown tachylytic band and is exposed on the top of Flow No. 5. Thickness of the band is 0.50 meter. [Fig. No.2.34]

FLOW NO.: 07

The flow is compact aphanitic basalt flow. The flow has developed closely spaced jointing pattern at the bottom and broadly spaced jointing pattern is developed in the top portion of the flow. Due to hydrothermal alteration the top portion has become reddish brown in colour. The flow consists of mainly three sets mutually perpendicular sets of joints in which two sets are vertical and one set is horizontal. Due to these three sets of joints compact aphanitic basalt is dissected in rectangular blocks. Due to percolation of water and its circulation around these joint blocks which at a time gives an appearance of conglomerates. The extent of weathering is very high producing thin weathered skin around the spheroids. [Fig.No.3.35] The flow appears from M.S.L. 695.73 m to M.S.L. 719.50 m i.e. having thickness of 23.77 m.

This flow is exposed at the left side along the ghat section from M.S.L. 719.50 m. It is compact aphanitic basalt flow. The top of the flow has broadly spaced jointing pattern whereas the middle and lower portion of the flow show closely spaced jointing pattern. [Fig.No.2.36] The thickness of the flow is 22.87 meters.

FLOW NO.: 09

The flow starts appearing from M.S.L. 742.37 m. It is hard compact aphanitic basalt flow. The top portion has developed closely spaced jointing pattern and the bottom portion has developed medium to broadly spaced jointing pattern. [Fig. No. 2.37] The flow has thickness of 6.10m.

FLOW NO.: 10

This is the top most flow present in Sautada ghat section starts appearing from M.S.L. 748.47 m. It is black compact aphanitic basalt flow in which three mutually perpendicular sets of joints are prominently seen where two sets are of vertical joints and one set is of horizontal joints. Due to broad spacing in the joints the compact aphanitic basalt is dissected into huge rectangular blocks. [Fig. No. 2.38] Closely spaced jointing pattern is seen in bottom portion at many places. Its top portion is exposed on plateau portion of the hill and is covered by black cotton soil due to intense weathering. The thickness of top flow is 29 m. at M.S.L. 777.47

Dyke

A dyke is exposed in middle portion of Flow No.SW 8. The dyke is having thickness of 1m. The trend of the dyke is N 10^0 W-S 10^0 E. It is black compact aphanitic basalt. [Fig. No.3.39] Jointing pattern in dyke is exposed in vertical cuts. Out of the two sets of joints, exposed at the surface, horizontal sets of joints are rather closely spaced, whereas vertical set of joint is broadly spaced. Due to these sets of joints big joint blocks are developed along the margin. The dyke occurs in weathered condition on both side and in the central portion the dyke rock is in fresh condition. The contact of the dyke with the country rock is clearly exposed along the road cut.

Lava Flows of Sautada Ghat Section

Fig. No. 2.31 Flow No. 1 Green Tachylyte

Fig. No. 2.32 Contact Between Flow No. 2 and Flow No. 3

Lava Flows of Sautada Ghat Section

Fig. No. 2.33 Contact between Flow No. 4 and Flow No. 5

Fig. No. 2.34 Flow No. 6 Red Tachylyte

Lava Flows of Sautada Ghat Section

Fig. No. 2.35 Flow No.7 Development of Spheroidal weathering

Fig. No. 2.36 Close Jointing Pattern in Flow No. 8

Lava Flows of Sautada Ghat Section

Fig. No. 2.37 Broadly Spaced Jointing Pattern in Flow No. 9

Fif. No. 2.38 Three set of Jointing Pattern in Flow No. 10

Lava Flows of Sautada Ghat Section

Fig. No. 2.39 Dyke exposed in Sautada Ghat Section

Fig. No. 2.40 Distant View of Sautada Ghat

Traverses Taken of Garmatha Ghat

2.4 Geology of Garmatha Ghat Section

This ghat section is exposed on Beed-Ahmadangar road in the district of Beed in Marathwada region. The ghat section is available at a distance of 50 km. from township of Beed at village Dongarkinhi at base

M.S.L. 655 and ends at village Bhatepura situated at top of ghat section at M.S.L. 790 m. The ghat section comprises of six flows.

Out of six flows four flows are compact Aphanitic basalt flows, one flow is of red tachylyte and one flow is Amygdaloidal basalt flow.

FLOW NO.: 01

The flow starts appearing from Beed 50 km stone at M.S.L. 655.48 m. The flow is compact aphanitic basalt flow. It is exposed at Dongarkinhi Village near Nallah. The compact basalt flow is having very tight jointing pattern. [Fig. No.2.41] Due to tight jointing pattern the percolation of water is very difficult. The thickness of the flow is 8.52 m.

FLOW NO.: 02

This flow is exposed along road cut at the right side of road when one is facing towards Beed. The flow is compact Aphanitic Basalt flow. The flow has developed very closely spaced jointing pattern. [Fig. No.2.42] The middle portion has developed closely spaced jointing pattern whereas vertical joints are commonly seen on the top. The flow has thickness of 6 m (M.S.L. 664 m to M.S.L. 669 m)

FLOW NO.: 03

This is a band of red tachylyte between Flow No.2 and Flow No.4.

The thickness is approximately 0.50 meter. [Fig. No. 2.43]

FLOW NO.: 04

This flow is of thin irregular Amygdaloidal Basalt. The flow starts from M.S.L. 670.0 m to M.S.L. 701 m having thickness of 31 m. The flow has well developed sheet jointing pattern. [Fig. No.2.44]

FLOW NO.: 05

The flow starts appearing from M.S.L. 701 m. The flow is compact Aphanitic basalt flow. Spheroidal weathering is very commonly seen at its bottom portion and middle portion. Three sets of joints are well developed at its top portion. Its top is hydrothermally altered. [Fig. No.2.45] Thickness of the flow is 34 m.

FLOW NO.: 06

The flow starts appearing from M.S.L. 735 m. It is topmost flow present in the ghat section. The flow is compact Aphanitic basalt flow. The flow is highly weathered. Spheroidal weathering is commonly seen and vertical jointing pattern is predominant. Its top portion is having three sets of joints. [Fig. No.2.46] Thickness of the flow is 54 m It is commonly exposed at Bhatapura village its top ends at M.S.L. 790 m.

Lava Flows of Garmatha Ghat Section

Fig. No. 2.41 Closely Spaced Jointing Pattern in Flow No.1

Fig. No. 2.42 Closely Spaced Jointing Pattern in Flow No. 2

Lava Flows of Garmatha Ghat Section

Fig. No. 2.43 Flow No. 3 Red Tachylyte

Fig. No. 2.44 Flow No. 4 Amygdaloidal Basalt

Lava Flows of Garmatha Ghat Section

Fig. No. 2.45 Flow No. 5 Compact Aphanitic Basalt Flow

Fig. No. 2.46 Flow No. 6 Compact Basalt with Three Set of Joints

3

CHARACTERS OF BASALT FLOWS

The detailed field characters of the basalt flows exposed in the various ghat sections of study area can be grouped into three types.

1) Thick extensive compact basalt flows;

2) Thin irregular amygdaloidal basalt flows and

3) Bands of red tachylytic basalt flows.

Compact Basalt Flow:

They are almost tabular in form having considerable thickness and large lateral extent. Their thickness varies from 7 m to 30 m and can be traced upto the distance of 20 km. [Karamakar-1974, Gupte et.al. 1974, Kulkarni S.R. 1975, Gupte et.al. 1977, Marathe et.al.1980] Immediately below top surface of the flow upto some depth the rock is generally hydrothermally altered, vesicular amygdaloidal.

Middle and lower portions of compact basalt flows are black in colour, nonvesicular, aphanitic or porphyritic in nature and are dissected in condition due to development of joints which are the contraction cracks formed during cooling and solidification of hot lava. Different types of jointing patterns can be observed in middle and lower portions of compact basalt flows.

Demarcation of two successive flows of compact basalts

Generally all the compact basalt flows exhibit various shades of black colour with similarity in lithological characters however, they can be separated from each other on the basis of following criteria.

1. Near the top of compact basalt flow same portion becomes vesicular, amygdaloidal and purple due to hydrothermal alteration. Thus two compact basalt flows can be demarcated from each other.
2. Lithologically two flows of compact basalt may be alike but sometimes there is vast difference in their jointing pattern, one flow may be having broadly spaced jointing where as another may have closely spaced or inconsistent jointing.
3. On the basis of appearance of weathered surface of joint blocks, two flows can be demarcated from each other. The compact aphanitic basalt always

develop smooth surfaces; Whereas porphyritic basalts develop rough and pitted surfaces due to weathering process due to dislodging of plagioclase phenocrysts.

Field characters of thick extensive compact basalt flows:-

Based on their megascopic characters compact basalt flows are divided into following categories.

i) Compact aphanitic basalt.

ii) Compact porphyritic basalt with small to medium sized plagioclase phenocryst and

iii) Giant phenocryst basalt with large sized phenocrysts having length more than 7.5c.m.

Variation in the thickness of compact basalt flows.

On the basis of thickness measured in field and particularly the L section it is observed that the thickness of compact basalt flows varied from ghat section to ghat section. The description of all types of lava flows in each ghat section is given in Table No.3.1

Table No: 3.1

Flow Description of Imampur Ghat Section.

Flow Type	Total No. of Flow	Maximum Thicknessin meter	Minimum Thicknessin meter	Average Thickness in meter	% of FlowType in Entire Ghat Section
Amygdaloidal Basalt	9	3 meter	0.5 meter	1.71 meter	15.25%
Compact AphaniticBasalt	4	6 meter	2 meter	3.62 meter	12.28%
Red Tachylyte	4	0.5 meter	0.5 meter	0.5 meter	2.00%
Compact PorphyriticBasalt	15	40 meter	1 meter	5.10 meter	70.47%

No. of Flows:-32
Total Thickness / Height of Ghat section :- 118 m.
(From M.S.L. 605 m to M.S.L. 723 m)

Table No: 3.2
Flow Description of Vambhori / Dongargan Ghat Section.

Flow Type	**Total No. of Flow**	**Maximum Thicknessin meter**	**Minimum Thicknessin meter**	**Average Thicknessin meter**	**% of Flow Type in Entire Ghat Section**
Amygdaloidal Basalt	1	2.82 meter	2.82 meter	2.82 meter	2.64%
Compact AphaniticBasalt	2	8.30 meter	7.00 meter	7.65 meter	14.36%
Compact PorphyriticBasalt	9	27.00 meter	2.75 meter	9.82 meter	83.00%

Total No. of Flows- 12
Total Thickness / Height of Ghat section:- 106.5 m.
(From M.S.L. 613 m to M.S.L. 719.50 m)

Table No. :3.3

Flow Description of Sautada Ghat Section

Flow Type	**Total No. of Flow**	**Maximum Thicknessin meter**	**Minimum Thicknessin meter**	**Average Thicknessin meter**	**% of Flow Type in Entire Ghat Section**
Amygdaloidal Basalt	2	8.00 meter	2.00 meter	5.11 meter	6.82%
Red Tachylyte	2	0.57 meter	0.50 meter	0.53 meter	0.66%
AphaniticBasalt	6	49.70 meter	6.10 meter	23.33 meter	92.52%

Total No. of Flows :- 10

Total Thickness / Height of Ghat Section :-150 m.

(From M.S.L. 627.43 m to M.S.L. 777.43 m)

Table No: 3.4

Flow Description of Garmatha Ghat Section.

Flow Type	**Total No. of Flow**	**Maximum Thickness in meter**	**Minimum Thickness in meter**	**Average Thickness in meter**	**% of Flow Type in Entire Ghat Section**
Amygdaloidal Basalt	1	31 meter	31 meter	31 meter	23%
Red Tachylyte	1	0.5 meter	0.5 meter	0.5 meter	0.75%
Aphanitic Basalt	4	55 meter	6 meter	25.88 meter	76.25%

No. of Flows :- 6

Total Thickness / Height of the Ghat Section :- 135 m.

(From M.S.L. 655 m to M.S.L. 790 m)

From the above data it is clear that compact basalt flow are predominant in the four ghat sections of study area. However, compact porphyritic basalt flows are predominant in Vambhori / Dongargan and Imampur ghat sections where as in Sautada and Garmatha ghat sections compact aphanitic basalt flows are more predominant and are devoid of porphyritic basalt flows.

During the field work it was observed that there is definite variation in the field characters of compact basalt flows from top portion to the bottom portion of the flows, irrespective of megascopic characters i.e. whether it is of compact aphanitic basalt, compact porphyritic basalt or giant phenocryst basalt. The top portion of every flow is vesicular amygdaloidal. Due to hydrothermal alteration top portion of some flows have become purple. However, it was also observed that top portions of some flows are free from hydrothermal alteration.

The top portion of compact basalt, being vesicular amygdaloidal it is almost un-jointed in fresh conditions. However, on weathering sheet jointing is developed in it. [Flow No. V 6; Fig. No. 3.1]

The middle portion and bottom portion of each compact basalt flow is free from vesicles and amygdales are occurring in true sense as black compact basalt. The middle and top portions of the flows are always dissected by joints, which are contraction partings developed during cooling and consolidation of lava. These contraction points open up at the surface on exposure to atmospheric conditions. These flows have different patterns of jointing and joint spacing. It is also observed that in aphanitic flows there is some consistency in jointing pattern but such consistency was not noticed in compact porphyritic basalt and giant phenocryst basalt flows.

Some flows are dissected by mutually perpendicular three sets of joints [Flow No. SW10; Fig. No.3.2] due to which the joint blocks have acquired rectangular shape and are stable one above the other. In most of the cases vertical joints deviate from the vertical developing wedge shaped joint blocks. [Flow No.G2; Fig. No.3.3]

The jointing may be closely spaced having distance between consecutive joints of 0.15 m [Flow No.V2; Fig. No.3.3] or broadly spaced [Flow No.V1; Fig. No.3.9] having distance between consecutive joints more than 1 meter. It is also observed that in closely spaced jointed flow deep weathering has taken place due to percolation of water. Whereas in broadly spaced jointed flow spheroidal weathering is developed due to mutually perpendicular sets of joints. [Fig. No. 3.8]

The weathered surface of compact aphanitic basalt have smooth appearance. [Flow No. SW8; Fig. No.3.10]

Whereas the weathered surface of porphyritic basalt and particularly in giant phenocryst basalt have rough and pitted appperace due to failing away of plagioclase

Fig 3.1: Hydrothermal Alteration of Flow No.6 of Vambhori/ Dongargan Ghat Section

Fig 3.2: Flow No.10 In Sautada Ghat Section Showing 3 Sets of Joints.

Fig 3.3: Flow No.2 in Garmatha Ghat Section Showing Closely Spaced Joints and wedge shape.

Fig 3.4: Flow No.3 of Imampur Ghat Section with Rough Appearance Due to Falling Away of Plagioclase Phenocryst Due to Weathering.

Fig 3.5: Flow No. 1 Green Tachylyte in Sautada Ghat Section.

Fig 3.6: Flow No. 22 Red Tachylyte in Imampur Ghat Section.

Fig 3.7: Dyke exposed in Sautada Ghat Section.

Fig 3.8: Product of Spheroidal weathering in Garmatha Ghat Section.

Fig 3.9: Broadly Spaced Jointing Pattern in Lava Flows of Garmatha Ghat Section.

Fig 3.10: Wedge shaped Jointing Pattern in Lava Flows of Sautada Ghat Section

phenocryst on weathering in Imampur Ghat section is commonly seen. [Flow No.3; Fig. No.3.4]

As mentioned above the top portion of compact basalt flow is vesicular or amygdaloidal and is un-jointed. The middle and lower portion of every flow is compact and jointed in nature. However, there is wide variation of the proportion between the thickness of top amygdaloidal portion and compact basalt portion of a flow.

On an average for the total thickness of basalt flows occurring along all the ghat sections only 10% to 12% of the portion towards top of compact basalt flows are vesicular and amygdaloidal in nature where as 88% to 90% middle and lower portions of these flows are compact as well as jointed.

The critical observations of the contact surfaces of two successive flows were carried out. Flow contacts are clean without any sign of weathering in the top portion of underlying flows.

The contact between the flows was found to be fused one which is an indication of rapid eruption of successive lava flows one after the other without any time gap in between the two.

Amygdaloidal Basalt Flows:

These are compound flows occurring as thin irregular bulbous in form. They are having small thickness and limited lateral extent. Such small irregular flows are amygdaloidal throughout their thickness. Very rarely thick tabular flows of amygdaloidal basalt also occur. [Karmarkar 1974, Gupte et.al. 1974] Most of the empty gas cavities are partially filled or completely filled up with secondary minerals i.e. white coloured zeolites; silica calcite and green coloured chlorophaeite. Very rarely empty gas cavities may occur in these flows.

Demarcation between two successive amygdaloidal flows:

Thin and thick flows of amygdaloidal basalt occur one above the other forming a thick pile. As contacts between the flows are fused the entire pile of amygdaloidal basalt flows occur as one homogenous mass. As amygdaloidal basalt flows are amygdaloidal throughout their thickness and are un-jointed therefore it is rather difficult to demarcate one flow from the other. However following criteria are used to demarcate one flow from other.

1) Between two flows of amygdaloidal basalts usually there is [Karmarkar-1974] a line of separation with the help of which the flow can be demarcated. Quite commonly the line of separation is reddish due to hydrothermal alteration at the contacts. This line is often made very conspicuous by occurrence of groove formed by weathering and removal of hydrothermally altered material.

2) Pipe amygdales commonly occur at the bottom of the flows.

3) The flow tops are generally ropy and reddened.

4) In case of thicker amygdaloidal basalt flows sometimes in the lower portion of the flow the amygdales are scanty and jointing is developed in that portion. In such cases two successive thicker flows can be demarcated from amygdaloidal top portion of underlying flow.

5) In addition to the above criteria the colour difference due to varying of hydrothermal alteration, variation in size and frequency and nature infillings of gas cavities also indicate different flows.

Bands of Tachylytic Basalt:

The glassy basalt pitch black in colour formed due to sudden chilling of lava are called as black Tachylytic Basalt. But In nature as compare to these black tachylytic basalts, hydrothermally altered red or green tachylytic basalts are also common. The tachylytic basalts are hard and in sound condition when they are not exposed i.e. when they are covered by sufficient cover of basalt flows but on exposure these tachylytic basalts become soft and are termed as red and green boles respectively.

The tachylytic basalt generally occurs as thin discontinuous bands between two basalt flows and their thickness varies from 0.20 m to 0.5m. Even thick tachylytic basalt bands have limited lateral extent and they gradually taper and pinch out after some distance.

Field characters of basalt flows exposed in the ghat sections of the study area show the following types of basalt flows.

1) Thick extensive compact aphanitic basalt flows.
2) Thick extensive compact porphyritic basalt flows.
3) Thin and thick irregular amygdaloidal basalt flows.
4) Discontinuous lenses and thin bands of red / green tachylytic basalt flows.

Tachylytic Basalts:

During the field work we came across two types of tachylitic basalt

- Green Tachylytic Basalt
- Red Tachylytic Basalt

1) **Green Tachylytic Basalt:-** Thin band of green tachylytic basalt occurring in Sautada ghat sections. This band is occurring at the base. It has thickness of 0.57 m. [Flow No.1; Fig. No.3.5]
2) **Red Tachylytic Basalt:-** Bands of red tachylytic basalt are commonly seen in Imampur ghat section where there are four flows of red tachylytic basalt are exposed. Each flow is clearly

visible and has thickness ranging from 0.15m to 0.75m. Flow No. 2;22;29 and 31 respectively of Imampur ghat section are the bands of Red Tachylytic Basalt. [Flow No.22; Fig. No.3.6]

One flow of Red Tachylytic Basalt is exposed in Sautada ghat section, Flow No. 6 having thickness of 0.50 m is separating Flow No.5 and Flow No.7.

At Garmatha ghat section only one Flow No.3 is exposed of Red Tachylytic basalt having thickness of 0.45m separating Flow No.1 and Flow No3. [Fig. No.3.44] However, Vambhori ghat section is devoid of red tachylytic basalt flow.

In fresh condition where red tachylytic basalt flow is not exposed to the atmospheric conditions, it occurs as a sound rock having very much closely spaced jointing, However, it is highly susceptible to weathering and occurs in the form of friable material.

Field Characters of Dyke :-

One dyke having thickness of one meter is exposed in Sautada ghat section. The trend of the dyke is N 10^0 W-S 10^0 E. It is black aphanitic basalt. Jointing in the dyke is clearly exposed in vertical cuts. Both the sets of joints are closely spaced. [Fig.No.3.7] The dyke occurs in weathered condition. The contact between dyke and country rock is clearly exposed along road side.

4

PETROGRAPHY OF BASALT FLOWS

Petrographic study is documentation of the observed facts about a rock, in terms of mineralogy and texture. This study is of great importance in elucidating the genetic history of an igneous rock because it throws light on the mineralogical paragenesis and the cooling behavior of the rock as reflected in its texture.

The chemistry of a rock alone cannot be the sole parameter to depict the complete character of the rock and to discriminate it from another with similar composition but formed in different cooling environment. Though in recent years much emphasis is being laid on major and trace element chemistry of rocks to interpret their genetic history, the petrography is yet felt extremely necessary, for petrographic interpretation and to recognize subtle differences in crystallization history of the same magma under diverse conditions.

Such discriminations may be necessary in course of field studies of igneous bodies and therefore, acquaintance with the petrographic techniques is essential for petrological investigations. Thus petrography is the ready means of communicating the physical appearance and other observable features of a rock. Though the description of rock starts in the field but much can't be achieved particularly about the mineralogy if the rocks are Aphanitic i.e. where the grains cannot be clearly resolved even under hand lens. Therefore, field descriptions or megascopic characters need to be supplemented by mineralogical and textural details recognized in thin sections of rocks examined under petrological microscope.

As mentioned above Petrographic study throws light on-

1) Physical conditions under which an igneous rock consolidated from magma. The grain size and some other textural features are directly related to the rate of magma cooling.
2) Offers clue to establish the order of crystallization of the constituent minerals.
3) Helps to understand the kinetics of crystal growth, possible reactions during crystallization, disequilibrium relations and controls on mineral alteration etc.
4) Helps to recognize flowage characters of the crystallizing magma.

5) Helps to assess the cooling history of the rock after consolidation i.e. sub-solidus cooling of the rock under favorable situation from some Intragrain texture.

6) Indicates probable temperature level, broad compositional character and nature of volatile constituent of magma.

PETROGRAPHY OF STUDY AREA.

As mentioned earlier 21 flows have been demarcated form various ghat sections of the study area. One flow is of dyke occurring in Sautada Ghat and rest twenty are of lava flows from all ghat sections. The division of flows according to ghat sections for study area has been given in Table No.4.1.

Table No: 4.1

Summary of Phyric and Aphyric Lava Flows of Study Area

Sr.No.	Name of Ghat Section	Abbreviation Used	No. of Flows Studie d	Phyric Flow	Aphyric Flow
1	Imampur Ghat	I	8	7	1
2	Vambhori Ghat	V	4	2	2
3	Sautada Ghat	S W	6	5	1
4	Garmatha Ghat	G	3	1	2
		Total	**21**	**15**	**6**

A) **SUB DIVISION OF PHYRIC FLOW:-**

1) **Plagioclase Phyric Basalt Flow:-**These are Fine grained, Compact Basalt massive in nature with Plagioclase laths as Phenocryst. [Flow No. V1; V11; SW8; I6; I30; I32].

2) **Plagioclase Clinopyroxene Phyric Basalt Flow:-** They are medium to coarse grained made up of clusters of Plagioclase and Clinopyroxenes exhibiting Glomeroporphyritic textures.

[Flow No: SW3; SW9; SW10] of Sautada Ghat section and [Flow No. I3, I8, I18] of Imampur Ghat section.

B) **Subdivision of Aphyric Flow:-**

1) **Aphyric With Microphenocryst of Plagioclase Flow::-**

It is fine grained massive compact basalt with minute presence of microphenocryst of Plagioclase showing sub-ophitic texture of Clinopyroxene in groundmass.[Flow No.V1 and Flow No. I20]

2) **Aphyric with Mafic Microphyric flow:-**

It is medium grained containing Microphenocryst of Clinopyroxene. [Flow No.G1; G2; V12]. One flow of Dyke in Sautada Ghat section also falls in same group.

Description of Mineral assemblage in study area:-

The basaltic lava flows can be differentiated into primary and secondary constituents. In general mineralogy of any rock can be explained in terms of essential minerals and accessory minerals present in the rock. Though the Deccan Trap Basalt appears to be remarkably uniform in nature in megascopic studies but in microscopic studies it can be differentiated into medium grained; coarse grained and fine grained having different textural characters. In the petrological study of rock samples of study area the mineral constituents have been classified into major and minor constituents. The major constituents are Plagioclase, Clinopyroxene, Olivine, Glass whereas the secondary minerals occur as vesicular infilling in lava flows, the minerals are various types of Zeolite group minerals like, Stilbite, Natrolite, Apophylite, Chabazite etc. and

Silica group minerals such as Quartz; Agate, Chert; Flint; Jespar; Chalcedony etc. Calcite and secondary glass are also present.

Proportion of these minerals by volume occurring in Basalt flows in study area is given in Table No.4.2

Table No:-4.2 Proportion of Minerals By Volume[By Delesess-Rosiwaal Method]

Name of Mineral	Phyric Flow	Aphyric Flow
Plagioclase	46.50%	40.50%
Pyroxene	9.17%	10.00%
Olivine	15.50%	3.50%
Iron ore	8.23%	10.00%
Glass	20.00%	33.00%
Amygdales	-	3.00%

Primary Constituents

Plagioclase:-

Plagioclase is the prominent and large sized lath-shaped mineral occurring in Basalt flows and its percentage by volume is 46.50% [Phyric Flows] and 40.50% [Aphyric Flows] the average of two being 43.50% In Basaltic lava flows Plagioclase occurs in two generations. Plagioclase laths of first generation are occurring as Phenocrysts, whereas slender laths of plagioclase of second generation are occurring in groundmass. The percentage of Plagioclase of first and second generation, occurring in basalt flows by volume is given in Table No: 4.3

Table No:- 4.3

Sizes of Plagioclase of First and Second Generations

Sr. No.	Flows	Flow No.	Length of First Generation (in mm)			Length of Second Generation (in mm)		
			Maximum	Minimum	Average	Maximum	Minimum	Average
1	**Pl. CPX Phyric**	I6; I30: I18	40	26	33	4	2	3.0
2	**Pl. CPX Aphyric**	SW5 SW8	13	7.5	10.25	4	2	3.0
3	**Pl. Phyric**	V2 : V11	85	80	82.5	3	2	2.5
4	**Aphyric With P1**	G5	12	7	9.5	2	0.5	1.25

Table No. 4.4

Sizes of Clinopyroxene Phenocryst of First and Second Generation

Flows	Flow No.	Length of First Generation (in mm)			Length of Second Generation (in mm)		
		Maximum	Minimum	Average	Maximum	Minimum	Average
Pl. CPX Phyric	SW3; SW9 SW 10	8.0	2.5	5.25	3.0	1.2	2.1
	I1; I3; I8; I18	10.0	2.0	6.0	2.0	0.2	1.1
Aphyric with Pl. microphenocryst	V1; V2	6	3	4.5	3	2	2.5
	I20	4	3	3.5	1	0.5	0.75
Aphyric with mafic Microphenocrysts	G1 G2	7.5	2.5	5.0	2	1	1.5
	V12	3	2	2.5	1.5	1	1.25

It is observed that large sized Phenocrysts occur in basalt having fine grained ground mass whereas slender laths of plagioclase occur in basalt flows having Aphanitic groundmass. Slender laths of plagioclase of first generation occur in Aphyric basalt flow [Flow No. G1; Fig. No. 4.1] and very large sized Plagioclase Phenocrysts occur in Phyric Flows [Flow No. SW5, Fig. No. 4.6].

Alteration:-

Most slender Plagioclase laths occurring in groundmass and large sized laths occurring as phenocrysts are very much fresh.

Corroded Border:-

Large sized plagioclase phenocrysts of first generation usually show corroded borders indicating resorption during eruption of lava. [Flow No.G1; Fig No. 4.2] Some phenocrysts show advanced stage of resorption, obliterating its original shape, the number of Plagioclase laths with corroded borders are more in Phyric flow [Flow No.G2; Fig No. 4.2: Fig No. 4.3].

Undulose Extinction:-

In Phyric Flows undulose extinction is very rare.

Inclusions:-

Inclusions of anhedral Pyroxenes and dust of Iron are very common in large sized Phenocrysts occurring in Phyric flows. Inclusions of olivine are almost absent in Plagioclase laths. Large number of workers have pointed out the inclusions in plagioclase in basaltic rocks. Inclusions of Augite in plagioclase are considered as rare in the Bhusawal Flows [Fermmor 1925 P.117] and in Satpura Basalts.

[Crookshank 1936 P.313, Benson [1944 P.73] remarked that the large feldspars are usually free from inclusions, but in medium grained rocks they often show inclusions of pyroxenes.

Twinning:-

Polysynthetic twinning without exception is common in both Plagioclase laths of first and second generation. However, polysynthetic twinning in Albite and complex Albite is very common in Plagioclase of first generation showing Carlsbad law, Sometimes in some phenocrysts fine strips of Pericline twinning is also seen crossing the Albite lamellae which are almost at right angles. However, it is not observed in the study area. Baveno twinning and Pericline twinning is not observed in the basalt flows of study area however Phenocryst of Plagioclase showing Albite twins is observed in Flow No G2 and Carlsbad twinning is seen in Flow No. SW5 [Fig. No.4.4].

Cross Twinning:-

Cross twinning of elongated laths of Plagioclase occurring as Phenocrysts and occurring as small laths in groundmass, is common feature. Generally two long laths of Plagioclase penetrate each other at right angles and rarely do they also make acute angle. [Flow No. SW8; Fig. No.4.5]

Cross twinning in Plagioclase laths appears to be a common phenomenon in Basalt flows in Deccan Trap area as reported by Machmohan 1883, Fermor 1925; Gorai 1951; Suryanarayan 1955;

Agashe 1956; Prasad 1959; Karmarker 1974; S. R. Kulkarni 1975; Marathe 1976; P. S. kulkarni, 1984; D. G. Chande 1987; A. V. Tejankar 2002; S. D. Doke 2013.

Gorai (1950) noted higher frequency of cross twins above Anorthite 50% Vishwanathan (1971 P. 475) observed cross twinning in Plagioclase with Anorthite content ranging between 30% to 50% Hence he concluded that there is no relation of cross twinning and chemical composition but it depends upon physical conditions like stress and strain and fluidity of magma or both. However, in Deccan Trap cross twinning has also been reported in Plagioclase with Anorthite content ranging from 55% to 70% also.

In the study area cross twinning is observed in Phyric Flows only. Flow No. SW8 and SW9. In which Anorthite content is 44.3% and 49.1% respectively. [Fig. No. 4.5]

Zoning:-

Zoning in Plagioclase Phenocryst is found to be well developed in Flow No. G5 which is Aphyric flow. The zoning in Plagioclase results from variation in the chemical composition during crystal growth and may be of different kinds, the commonest is Normal Zoning which involves a gradation from 'Ca' rich core to 'Na' rich outer layer. [Flow No. SW1 and Flow No.G5 respectively; Fig. No. 4.7] Yet another zoning in which the composition has changed rhythmically and repeatedly causing Rhythmic or Oscillary Zoning. [Flow No.V2; Fig. No. 4.9, Fig. No. 4.10] Less frequently the converse is true and that is called Reverse Zoning. [Flow No.V11; Fig. No.4.8].

Anorthite Content:-

Anorthile content of medium and large sized Phenocryst of Plagioclase was carried out by Michel Levy method. It is shown in Table No.4.5.

Table No. 4.5

Variation of Anorthite Content in Plagioclase by Michel LevyMethod

Plagioclase Phyric Flow	Sr. No.	Flow No.	Anorthite Content
	1	V11	45%
	2	G5	60%
	3	SW5	45%
	4	SW8	60%
	5	SW9	60%
Plagioclase + CPXphyric Flow	6	SW10	68%
	7	I1	45%
	8	I3	45%
	9	I8	45%
	10	I18	45%
Aphyric flow with plagioclase micro phenocryst	11	V2	45%
	12	G1	45%

From the above table Anorthite content of Plagioclase in Basalt Flows varies from 45% to 68% [Average being 50.66%]

The Literature surveys also shows that An content determined by other workers also falls between 50% and 70%, only in rare cases it cross 75%,Washington [1922] 66%, Fermor [1925] 65%, West [1957] 56%. According to Washington [1922] in Deccan Trap Basalts average An content is 66% Fermor [1925 p. 115] stated the Anorthite content to be 65% West [1957] found that An content is from 55% to57% for the Lower Traps and from 60% to 68% for the Upper Traps. In general the ground mass Plagioclase is more calcic in the Upper Traps. Thus the Plagioclase is more calcic in the Upper Traps. Thus the Plagioclase Phenocryst of Deccan Trap Basalts have average Anorthite content 64% and those of ground mass have average Anorthite content 55%.

It appears from the table that Plagioclase varies from 45% to 60% and the average Anorthite content in Phyric Flows is 51.8% where as it is 45% in Aphyric Flows.

Pyroxene:-

Generally well developed octahedral crystals of Augite are seen in chilled or rapidly cooled basalts only. Teall [1884 p.233], Fermor [1925 p.116] Crookshank [1936 p.302]. Occurrence of Aguite as Phenocryst is common in Basalt flows Barth [1931 p.205], Fermor [1925 p.116] stated that Augite is idiomorphic in glassy basalts, but it becomes sub idiomorphic to granular in coarsely crystalline basalt. In the Basalt flows of the area only Cinopyroxenes are present in all the Basalts. It is next to Plagioclase in abundance. It is colorless and nonpleochroic having R.I.β1.70.

The average proportion of pyroxene by volume for flows is 9.7% and in Plagioclase Phyric Flow is 10%.

Normally Pyroxenes present in flows are of two generations. At some places it occurs as granules from very fine granules in groundmass to small crystals interstitial to Plagioclase which is seen in Flow No. G1. [Fig.No.4.13]

Occurrence of Pyroxene as unhederal phenocryst is very rare in basalt flows. Occasionally elongated slender prism of Augite are seen, Augite phenocrysts are much smaller than Plagioclase Phenocryst in size, however, in some flows [Flow No.V11; I6; I18; SW8; G1] euhedral phenocrysts of Pyroxene are seen showing one or two sets of cleavages. [Fig. No. 4.13] The maximum size of Augite Phenocryst of first generation is 21 mm. in Flow No. SW8 showing one set of cleavage. The average size of first generation of Augite Phenocryst is 4.5 mm in the study area ranging from maximum size of 21 mm to minimum size of 0.7 mm. The maximum size of second generation occurring in ground mass is of 4.5 mm in Flow No. SW8 and minimum size is 0.1 mm and the average being 1.1 mm.

Pyroxencs mainly occur as unhedral granules or subhedral grains in the groundmass, in almost all the flow types. The small unhedral Phenocrysts associated with the granules are seen in Phyric Flows and subhedral grains are observed in Aphyric Flows. Pyroxene Phenocryst occurs singly or in cluster giving rise to glomeroporphyritic texture. [Fig. No. 4.25] However, in Phyric flows its frequency is more. [Flow No. G5] In general Pyroxenes don't show well developed cleavages. However one set of cleavages is found. [Flow No.G1; Fig. No. 4.14] and two sets of cleavages are found in the basal section. [Flow No. G1; Fig. No. 4.13]

Alteration:-

Alteration of Pyroxene in the flows is not common, however, Pyroxene altered to chlorite is observed. [Flow No. SW8; Fig. No. 4.15]

Twinning:-

Generally twinning is rare in pyroxene. While discussing origin of twinning in Augite Subbarao [1972] pointed out that intrusive dolerites frequently display twinning in Augite and is not common in Gabbro and Basalt. The twinning seems to be connected with conditions of crystallization of magma and the pressure exerted on crystals. During the course of injection of magma into a narrow fissure twinning in Augite may be developed.

Some pyroxenes show undulose extinction which is particularly observed in Phyric Flows. This phenomenon is attributed mostly to variable composition and to inconspicuous zoning or strain effect.

Olivine:-

Out of the twenty one lava flows of the study area

Olivine is observed in sixteen lava flows. Olivine crystals are prismatic and somewhat euhedral and in Aphyric lava [Flow No. V5] it is present as phenocryst, with some cracks developed in it. The maximum size of Olivine Phenocryst observed in the study area is 10 mm [Flow No.V1; Fig. No. 4.16] and minimum size is 1mm with average size of 3.5 mm. The occurrence of Olivine in Basalt rock of this area is in agreement with Kuno [1950], MacDonald and Katsura [1964]. They have pointed out that such occurrence is common in Tholeiitic Olivine Basalts. The average proportion of Olivine in Phyric Flows is 15.5% and in Aphyric flows is 3.5% .This indicates that Aphyric Flows are richer in Olivine as compare to Phyric Flow.

Alteration:-

Alteration of Olivine to Iddingsite and Serpentine is common in all sixteen flows in the study area. Alteration of Olivine to Iddingsite has also been reported by Ian baker and Stephen E. Haggerly [1967] in the basaltic lava flows of St. Helena. As reported by them it is post diuteric alteration product of Olivine which is formed when the lava temp. was below 140^0C. The term diuteric was introduced by Sederholm [1926] which defines 'Those that have taken place in direct continuation of consolidation of magma of rock itself'. As reported by M. J. Baker [1966] and Gay and Lemaitre [1961] the lddingsite in the study area can be grouped into.

A] Pleochroic Varieties and

B] Non Pleochroic Varieties.

The pleochroic varieties show high content of iron as reported by Smith [1959] and Brousse [1961] and it is due to post diuteric alteration. The Pleochroic varieties are [Flow No.V5; Fig. No 4.10] with size 20 mm and all pleochroic varieties are dark red and dark red brown in color.

The Non pleochroic varieties show fibrous structure with golden yellow ground mass [Flow No.V2; Fig No. 4.17] as reported by Ian Baker and Stephen E. Haggery [1967].

Occurrence of olivine in the form of inclusion in plagioclase and pyroxene phenocryst [Flow No. V5; Fig. No. 4.17] have already been reported by various workers. Farmer (1925), Kuno (1950), Subbarao (1972), Marathe (1976),Yerkes (1957), Karmarkar (1974), S.R. Kulkarni (1975). Such inclusions of Olivine in Plagioclase are observed in flow of this area. As reported by these workers the flows in this area also show development of Magnetic iron ore around olivine granules and along cracks.

Iron ores:-

Iron ores are the common accessory constituents of Basalts. They occur as rods, plates, needles, granules and dust [Flow No.V12: Fig. No. 4.19 to 4.21]. They also occur as microlites and crystallites in the glass. Dust of iron ore occurs in the glass imparting black color. Inclusions of dusty iron ore are frequently seen in Plagioclase Phenocrysts as in the Phyric Flow [Flow No. SW8; Fig. No. 4.20] and in Aphyric Flow. [Flow. No. V12; Fig. No. 4.19] Occurrence of iron ore is different in different type of flows of basalt i.e. Phyric Flow [Flow. No. SW8] and Aphyric Flow [Flow No.V12]. In the glassy portion of flows granules and needles are concentrated within the patch of primary glass as in case of Phyric Flow. [Flow. No. SW8]

However, in the well crystalline portion they appear as separate plates from the primary glass and their occurrence is in crystalline form as seen in Phyric flow [Flow.No. SW8; Fig. No.4.20] and in Aphyric Flow. [Flow No. V12; Fig. No 4.19]

Titanium ore:-

Next to iron ore titanium ore are commonly occurring accessary constituent of Basalt. The Most commonly occurring titanium mineral is Ilmenite. Presence of Titanium is seen in triangular form in Aphyric Flows. [Flow No G1; G2 and V12] The Ilmenite appears in black colored rhombohederal crystals however many crystals show alteration. Due to alteration the borders are corroded and there is change in rhombohederal form. The crystals are opaque, isotropic and are difficult to distinguish from iron ore. However, due to advanced stage of alteration titanium minerals mostly Leucoxene [Flow No. G1; Fig. No. 4.22] appears translucent and light brownish grey whereas trigonal symmetry is emphasized by thin parallel black bars crossing in three directions.

Occurrence of titanium in the form of Ilmenite has been reported by Harker [1909] Krishnamurthy and Cox [1977] Beane and Hooper [1988], Murari [1993] has reported presence of Magniesian Ilmenite of Kalsubai lava Flows.

TEXTURES

Microscopic study of thin sections of various Basalt flows of the rocks of the study area has shown almost all types of textural variations. The textures are Porphyritic, Poikilitic, Glomeroporphyritic, Intergranular, Intersertal, Ophitic, Subophitic, Flow texture.

Porphyritic Texture:-

When observed under microscope all types of flows show Porphyritic texture [Fig. No.4.1: Fig. No. 4.6] Both Phyric [Flow No. SW5; Fig. No. 4.8] and Aphyrictype basalt flows are having Plagioclase and Pyroxene as Phenocryst [Flow G1 Fig. No.4.1]

In all the flows groundmass consists of small laths of plagioclase granules of Pyroxenes iron ore, olivine granules and irregular patches of glass.

Plagioclase as Phenocryst:-

As stated above Plagioclase always occurs as a Phenocryst in Basalt Flows. However, size of phenocryst is different in various types of flows. In the Phyric flow of Vambhori / Dongargan Ghat section in lava flow V2 and V11 the maximum length of plagioclase Phenocryst is found to be 85 mm. (max) and 80mm.(max) with the width of 20 mm. and 18 mm. respectively. Whereas in Aphyric flows of Imampur Ghat Section in lava Flow No. I20 maximum length of plagioclase phenocryst is 4 mm. (max) and 3 mm. (min) with average length of 4.5 mm. respectively.

Zoning is observed in plagioclase phenocryst. i.e. Normal zoning [Flow No. G5; Fig. No. 4.9] and Oscillary zoning. [Flow No. V2; Fig. No. 4.10] Whereas Reverse zoning is also observed [Flow No. V.11; Fig. No. 4.8] Some phenocryst show resorption. [Flow. No. V2; V11 Fig. No. 4.11] The phenomenon of zoning and resorption of plagioclase Phenocryst is seen in phyric lava flows. In some phyric lava flows plagioclase phenocrysts are observed in clusters [Flow. No. SW5; Fig 4.6] Where as in some phyric flows [Flow No. SW8, G2; SW5; Fig. No. 4.3] show corroded borders. In Phyric flows inclusions of small granules of pyroxene are seen in plagioclase phenocryst [Flow No. G1; Fig. No. 4.1] well developed fresh laths of plagioclase phenocrysts are observed in Phyric Flows. [Flow No. V5; Fig. No. 4.23]

Pyroxene as Phsssenocryst:-

As compare to the size of plagioclase phenocrysts in lava flows of study area the pyroxenes which occur as phenocrysts are always smaller in size. The phenocrysts generally occur in cluster. They are unhederal and some show tendency to subhedral habit some show one set of cleavage. [Flow No. G1; Fig. No. 4.14] Twinning in Pyroxene is rare. In some flows it occurs in contact with Plagioclase phenocryst.

Glomeroporphyritic Texture:-

Clusters of Plagioclase phenocrysts are very common in Phyric Flows. [Flow No. SW5; Fig. No. 4.6] In some flows along with Plagioclase phenocrysts large number of Pyroxene and rarely olivine Phenocrysts are occurring giving rise to Glomeroporphyritic texture. The Glomeroporphyritic texture might have been formed due to simultaneous floating of crystals together. [Flow No.G5; Fig. No. 4.25]

Intergranular and Inter sertal Texture:-

In the Basalt flows sometimes Plagioclase phenocrysts get arranged in such a way that the triangular or polygonal interstices are formed. These interstices are occupied by small sized plagioclase laths, pyroxene granules, iron ore and altered olivine giving rise to Intergranular Textures. [Flow No. SW8; Fig. No. 4.29] When such interstices are filled by cryptocrystalline glassy material or secondary minerals like palagonite, chlorophaeite the texture is termed as Intersertal Texture. [Flow No. SW8; Fig. No. 4.30] Thus in the study area both Intergranular and Intersertal textures are observed in both Phyric and Aphyric flows.

Ophitic and Sub-ophitic Texture:-

When small laths of plagioclase are found to be enclosed in plates of pyroxene then the developed texture is called as Ophitic Texture. [Flow. No. V11; Fig. No. 4.28] In some flows along with plagioclase phenocrysts, large number of pyroxene and rarely olivine phenocrysts are occurring giving rise to Sub-ophitic texture. [Flow No.V12; Fig. No. 4.27] Generally Ophitic and Sub-ophitic texture is common in all types of flow.

Olivine As Phenocryst:-

In very few basaltic lava flows of study area. [Flow No. V1; Fig. No. 4.16] Olivine as a phenocrystis observed. The maximum grain size of 10 mm. Olivine as a phenocryst is observed in Flow No. SW9, SW10, V1, and V11. The phenocryst of Olivine is a well developed prismatic hexagonal shaped and it is having yellow to brown colour. The cleavage cracks are very common and are filled with iron oxide exhibiting dark brown colour. In some olivine grains alteration has proceeded mainly along the cracks. The common alteration product is fibrous Serpentine; Iron oxide thus, released during the alteration has gone to form magnetic iron ore in the neighborhood.

Flow Texture:-

The texture is observed in Aphyric Flow [Flow No. G2; Fig No. 4.26] in which small slender and elongated laths of plagioclase in groundmass are seen to have some linear orientation imparting the Rock Flow or Trachytic texture.

Vesicular and Amygdaloidal Texture:-

These textures are common in Basalt flows of Deccan Trap. Basalt being volcanic rock gas cavities of different sizes is very common. These gas cavities are generally filled up by secondary minerals like zeolites, calcite, silica, palagonite, chlorophaeites etc. giving rise to Amygdaloidal texture. When these gas cavities remain empty or are not filled by secondary minerals the texture is called Vesicular. Sometimes amygdales are irregular in shape with a rim of chlorophaeite or filled with zeolites. [Flow No. I1; Fig. No. 4.31 and Fig. No. 4.32]

Spherulitic Texture:

When viewed between the Crossed Nicols under the microscope the perfect black extinction cross appears and it remains constant when the stage is rotated through 360^0. In many lava flows this structure appears in bundles or group. [Flow No. G1; Fig. No. 4.36] Occurrences of Spherulites have been reported in acid volcanic or hypabyssal rocks but rarely their occurrence have been reported in basic lava flows.

The Spherulites show radiating growth of feldspar fibres only which indicates the viscous nature of lava. Spherulites are the products of devitrification of brown basaltic glass. Localization of Spherulites depends upon minute differences of physicochemical character, especially the content of water vapour and viscosity. [Flow No. G1; Fig. No. 4.34, Fig. No. 4.35, Fig. No. 4.36]

Sieve Texture:-

This texture is produced when a large plagioclase phenocryst is filled with many channel ways (holes) by secondary minerals. [Flow No. SW8; Fig. No. 4.21] This structure is formed when a phenocryst is densly charged with inclusions.

Swallow Tail Structure:-

The small crystals of plagioclase in a rapidly cooled magma due to their rapid growth and slow rate of diffusion in silicate melt grow with dendrite habit and are characterized by skeletal forms with considerable open space. This results in formation of structure which resembles a swallow tail. [Dyke; Fig. No. 4.33]

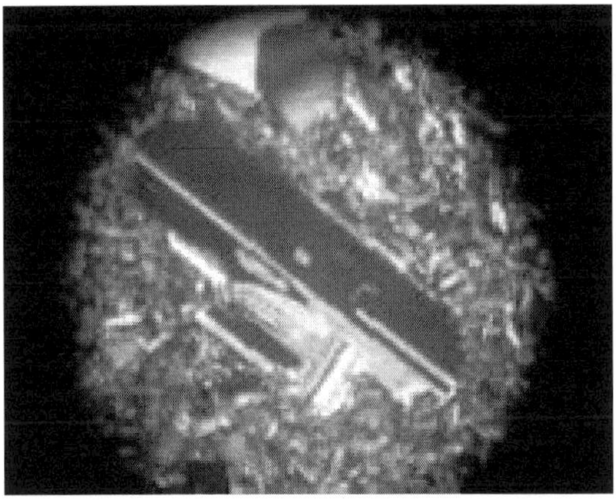

Fig. No: 4.1 Plagioclase Phenocryst In Flow No. G1 (Crossed Nicols 10x)

Fig. No: 4.2 Plagioclase Phenocryst with corroded border In Flow No : G2 (Crossed Nicols 10x)

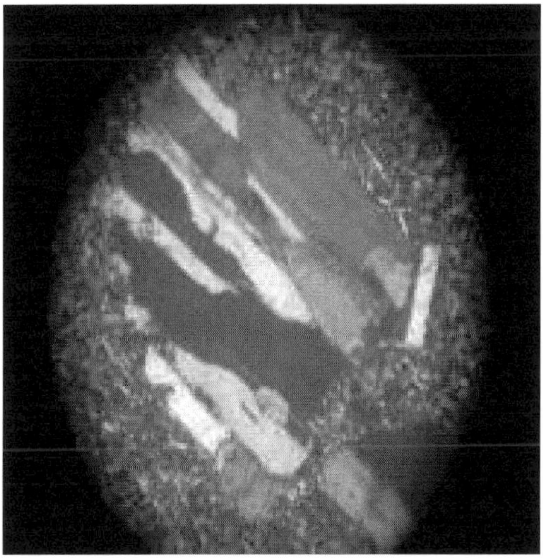

Fig. No: 4.3 Plagioclase Phenocryst with corroded border In Flow No : G2 (Crossed Nicols 5X)

Fig. No: 4.4 Plagioclase Phenocryst showing Carlsbad Twinning with corroded border In Flow No: SW 5 (Crossed Nicols 10X)

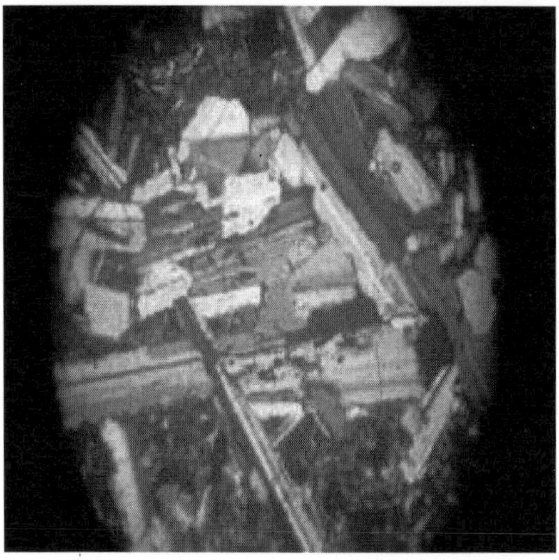

Fig. No: 4.5 Plagioclase Phenocryst showing Cross Twinning In Flow No: SW8 (Crossed Nicols10X)

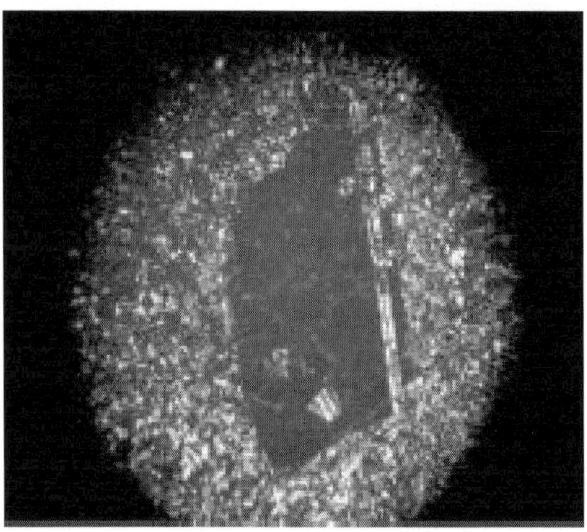

Fig. No: 4.6 Cluster of Plagioclase Phenocrysts In Flow No: SW5 (Crossed Nicols 10X)

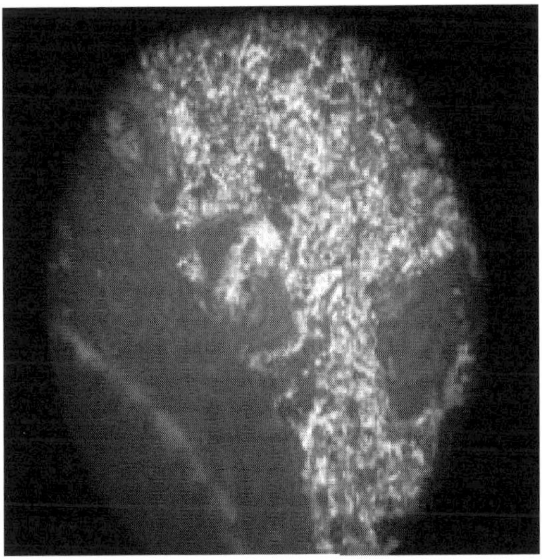

Fig. No: 4.7 Normal Zoning in Plagioclase In Flow No: SW1 (Crossed Nicols 10X)

Fig. No: 4.8 Reverse Zoning in Plagioclase In Flow No: V11 (Crossed Nicols 5X)

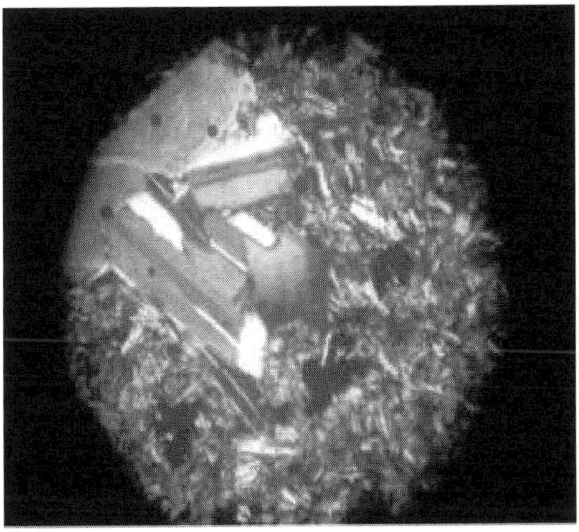

Fig. No: 4.9 Normal Zoning in Plagioclase In Flow No: G5 (Crossed Nicols 10X)

Fig. No: 4.10 Oscillary Zoning in Plagioclase with development of Iddingsite In Flow No: V2 (Crossed Nicols 10X)

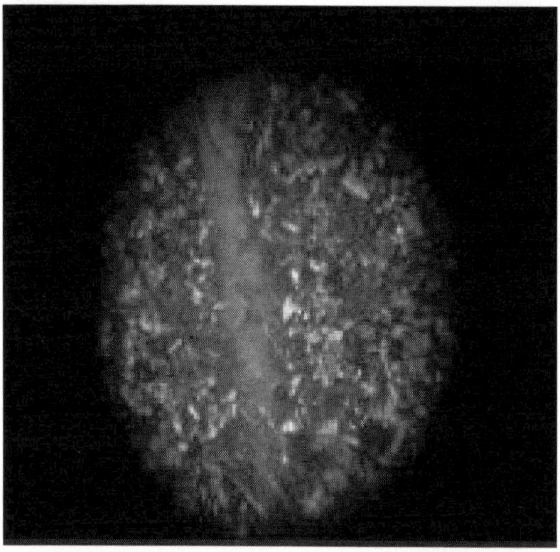

Fig. No: 4.11 Resorption of Plagioclase In Flow No: V2 (Crossed Nicols 10X)

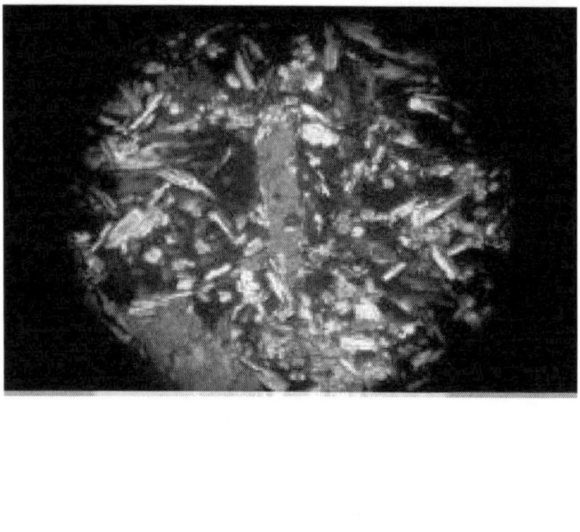

Fig. No: 4.12 Augite Phenocryst In Flow No. G1 (Crossed Nicols 10X)

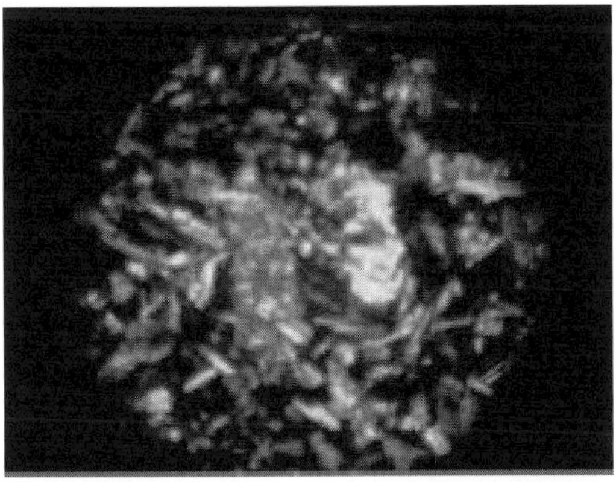

Fig. No: 4.13 Cluster of Augite Phenocryst having two set of cleavage In Flow No: G1 (Crossed Nicols 10X)

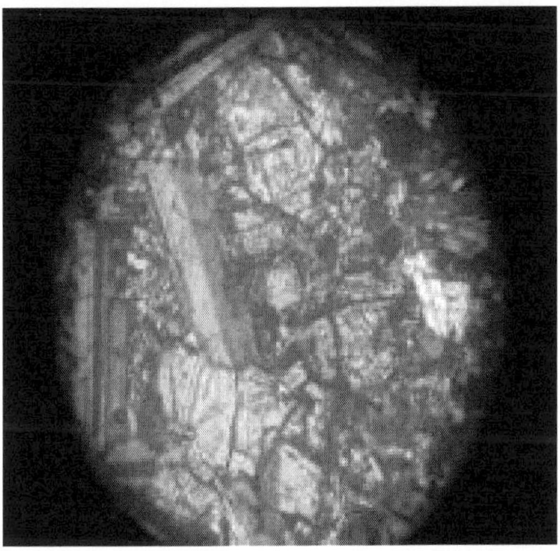

Fig. No: 4.14 Augite Phenocryst with one set of Cleavage In Flow No: G1 (Crossed Nicols 10X)

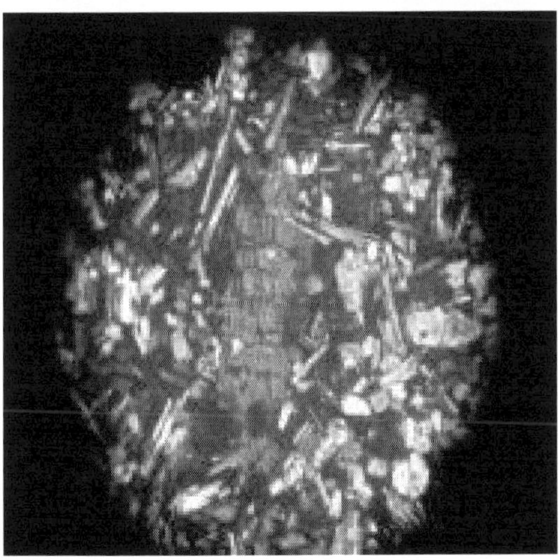

Fig. No: 4.15 Augite Altered to Chlorite In Flow No: SW8

(Crossed Nicols 10X)

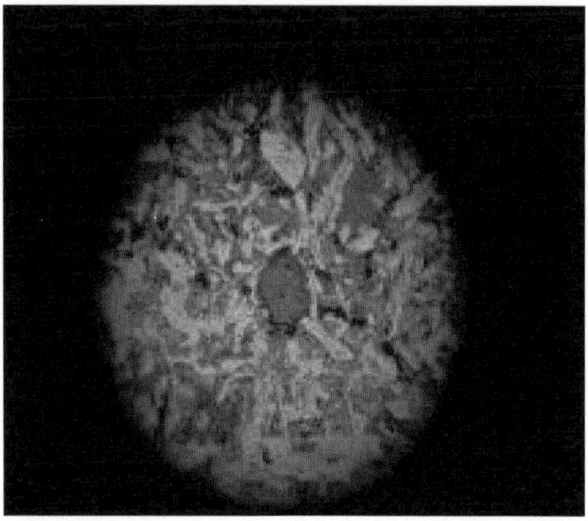

Fig. No: 4.16 Olivine Phenocryst In Flow No: V1 (PPL 10X)

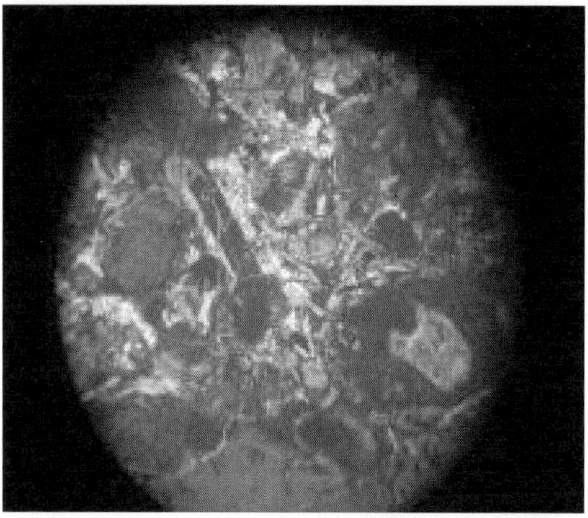

Fig. No: 4.17 Olivine In Cluster In Flow No: V5 (crossed Nicols 10X)

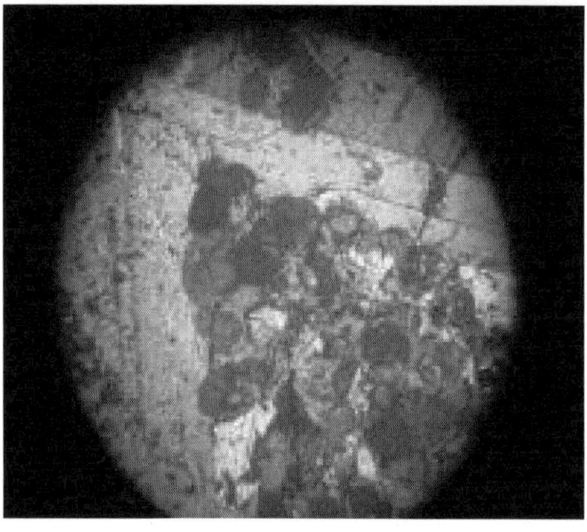

Fig .No: 4.18 Olivine Altered To Serpentine In Flow No: V2 (crossed Nicols 10X)

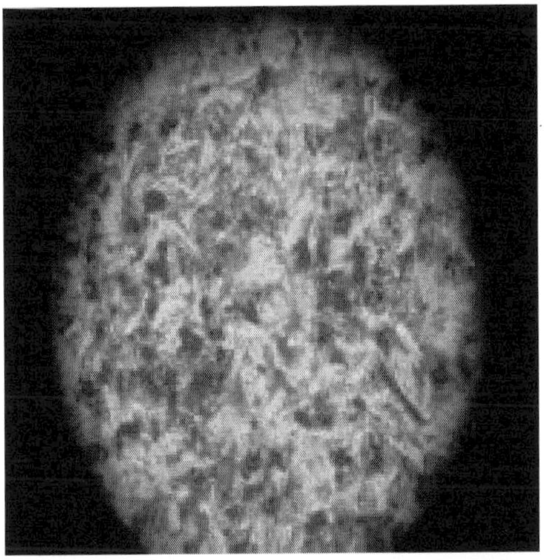

Fig .No: 4.19 Laths of Iron Ore in Flow No: V12 (PPL 10X)

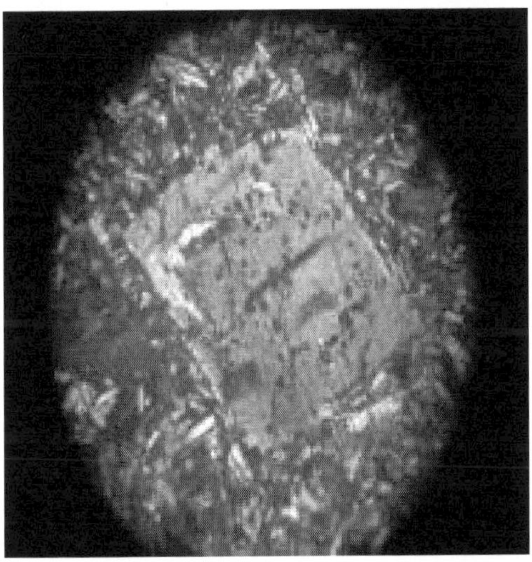

Fig .No: 4.20 Dust Particles of Iron Ore on Plagioclase Phenocryst In Flow W8 (Crossed Nicols 10X)

Fig .No: 4.21 Formation of Sieve Structure Due To Dust Particles of Iron Ore on Plagioclase Phenocryst In Flow No: SW8 (Crossed Nicols 10X)

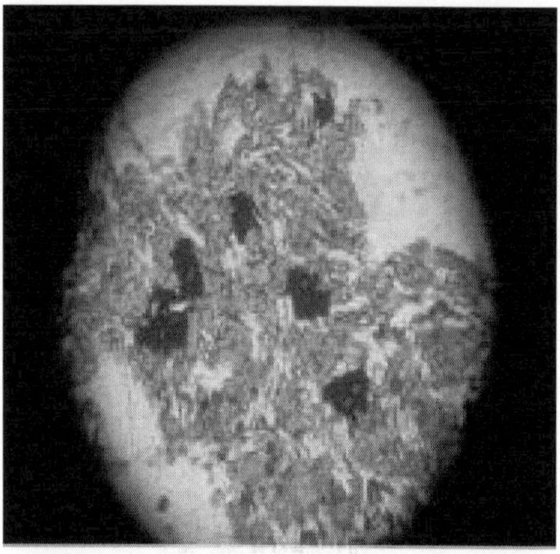

Fig .No: 4.22 Laths of Ilmenite In Flow No: G1 (PPL 10X)

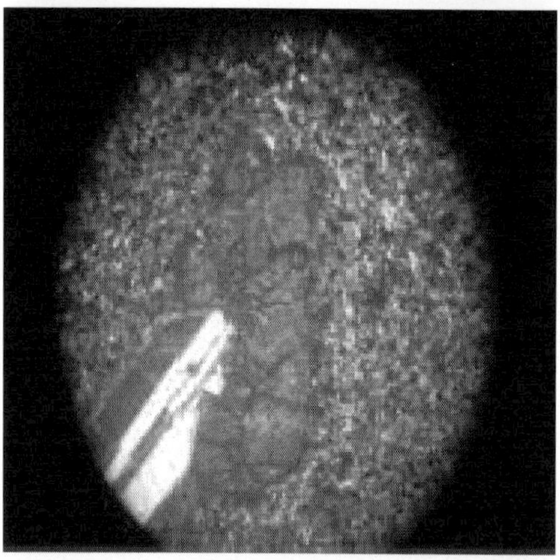

Fig. No: 4.23 Plagioclase and Olivine In Flow No: I6 (Crossed Nicols 10X)

Fig. No: 4.24 Intergranular Texture In Flow No: G5 (Crossed Nicols 10X)

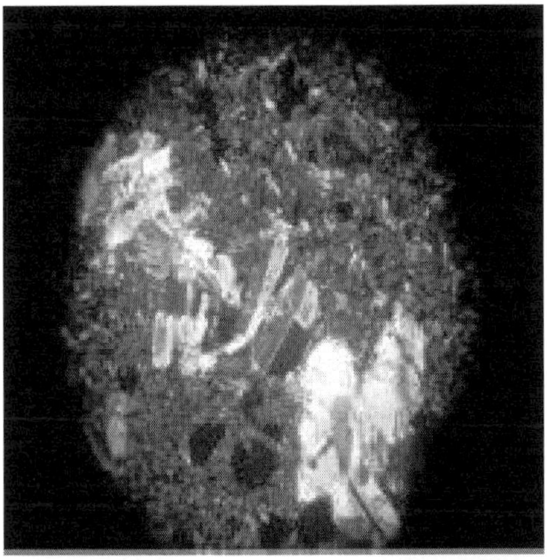

Fig. No: 4.25 Glomeroporphyritic Texture In Flow No: G5 (Crossed Nicols 10X)

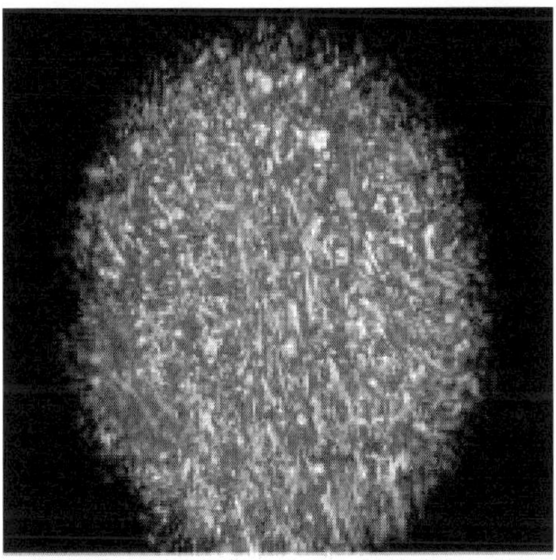

Fig. No: 4.26 Directive Texture In Flow No: G2 (Crossed Nicols 10X)

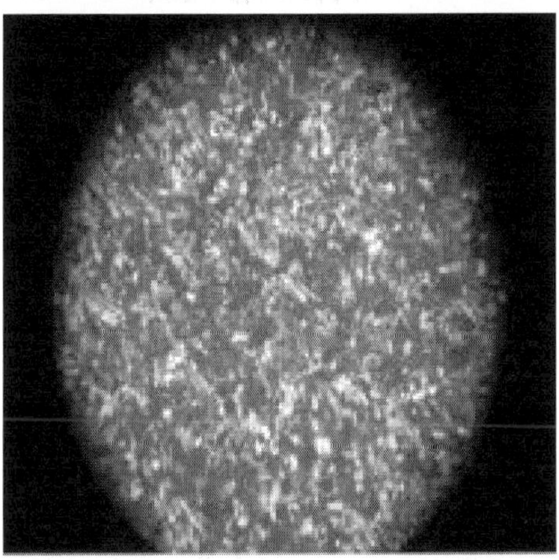

Fig. No: 4.27 Sub Ophitic Texture In Flow No: V12 (Crossed Nicols 10X)

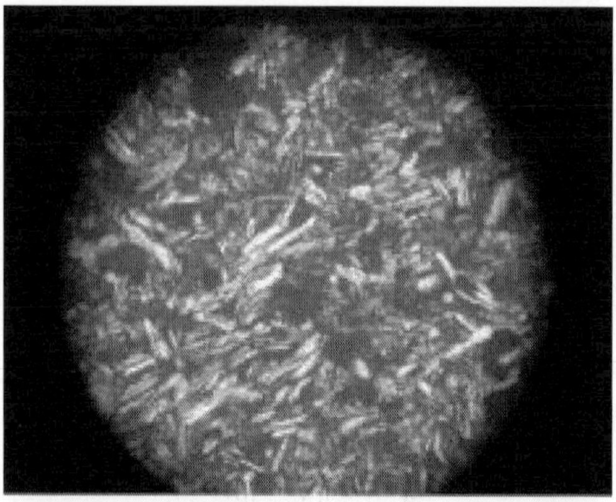

Fig. No: 4.28 Ophitic Texture In Flow No: V11 (Crossed Nicols 10X)

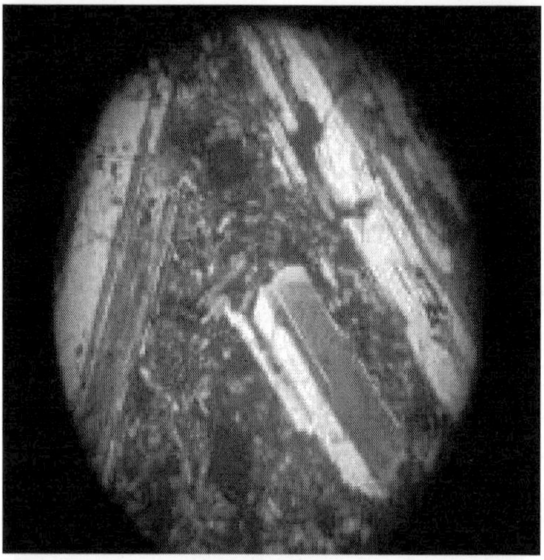

Fig. No: 4.29 Intergranular Texture In Flow No: SW8 (Crossed Nicols 10X)

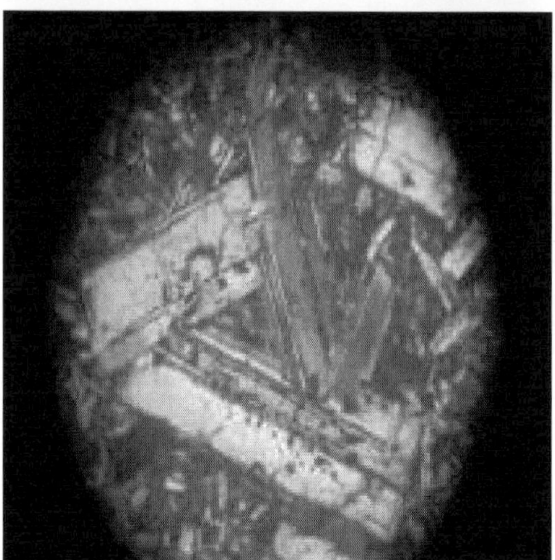

Fig. No: 4.30 Intersertal Texture In Flow No: SW8 (Crossed Nicols 10X)

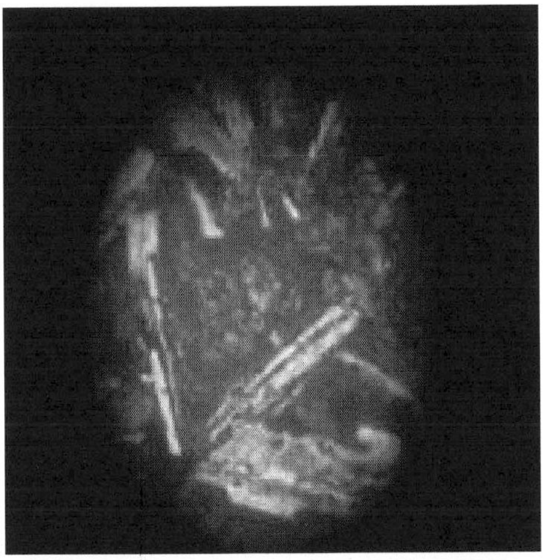

Fig. No: 4.31 Amygdaloidal Texture In Flow No: I1 (Crossed Nicols 10X)

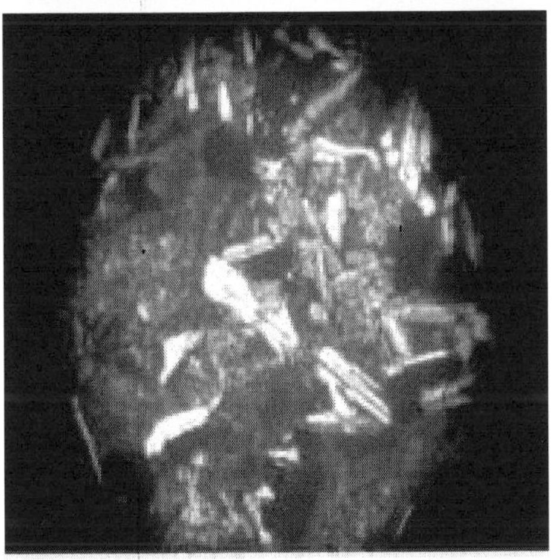

Fig. No: 4.32 Amygdaloidal Texture In Flow No: I1 (Crossed Nicols 10X)

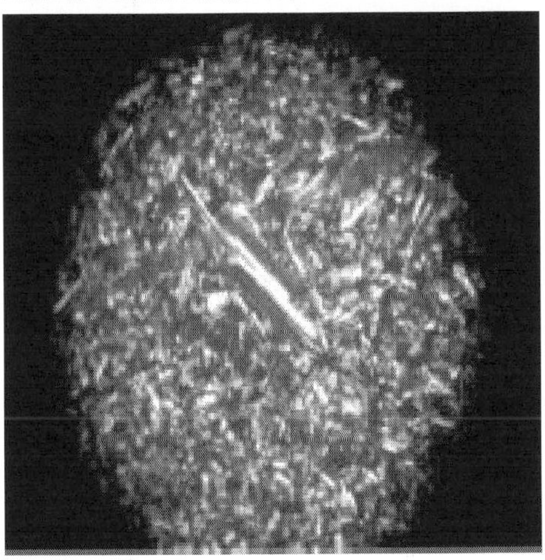

Fig. No: 4.33 Swallow Tail Texture In Dyke (Crossed Nicols 10X)

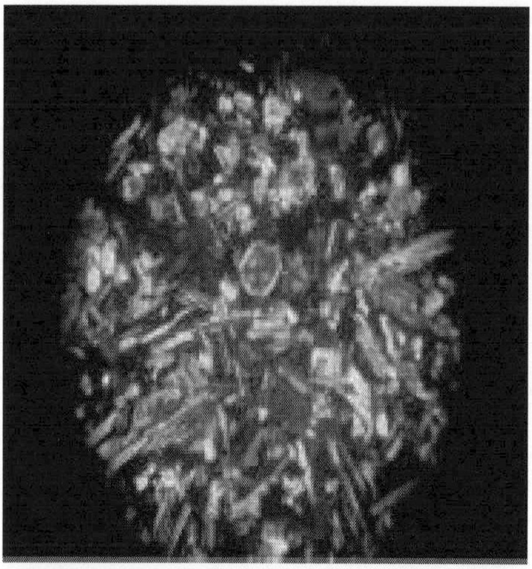

Fig. No: 4.34 Spherulitic Texture In Flow No: G1 (Crossed Nicols 10X)

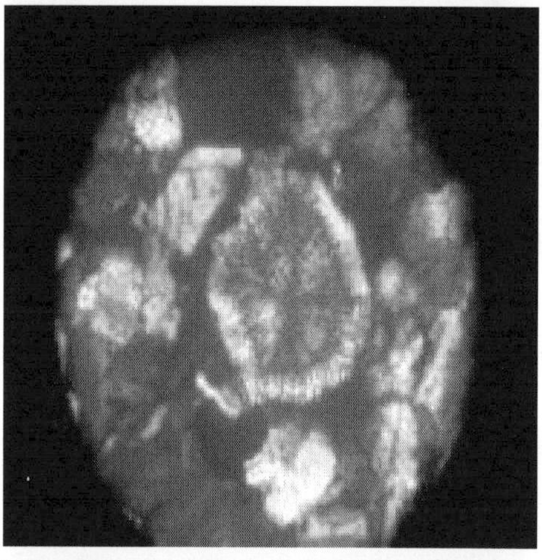

Fig. No: 4.35 Spherulite In Flow No: G1 (Crossed Nicols 40X)

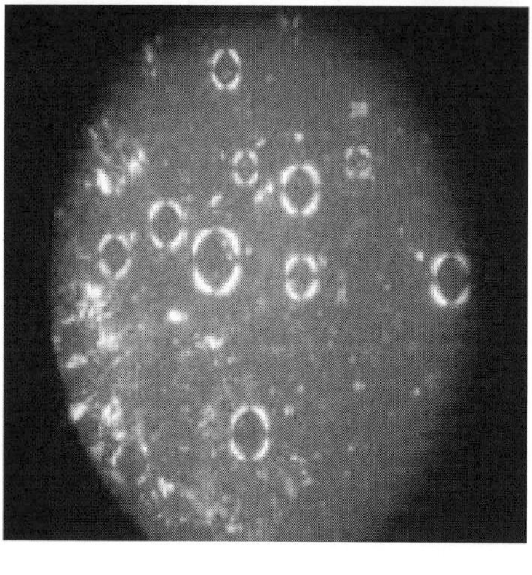

Fig. No: 4.36 Cluster of Spherulites In Flow No: G1 (Crossed Nicols 10X)

5

DISCUSSION AND CONCLUSION

The main object behind the present endeavour is to determine the nature of volcanicity chosen for the purpose. The field characters of the basalt in this regime throw appropriate light on the nature of volcanicity. The ghat sections provide ideal site for investigation as well as determination of field characters of the flows exposed there in. Accordingly traverses were taken which have been described in chapter III and chapter IV. In this way petrographic characters and chemical analytical data of in all 21 lava flows occurring in four such ghat sections have been described in chapter IV . In order to asses the nature of volcanicity and correlation of basalt flows, the interpretation based on these findings has been presented here. The field characters of the basaltic lava flows provide clues to the nature of volcanic activity.

As we know lava flows are formed by solidification of liquid which flows and spreads out in the form of a sheet. As compared to thickness these flows have large extent and thickness which depends on fluidity of the lava. With rise in fluidity and mobility of lava flows the ratio of lateral extent/thickness of flow [l/t] increases and it forms extended sheet after cooling and consolidation. Plain top and bottom pattern of lava flow is an indication of an undisturbed succession, forming almost horizontal lava flows exhibiting tabular form. With ample supply of lava the sheet attain considerable thickness covering extensive areas as well.

On the other hand if the lava is not fluid but is viscous in nature then it will tend to pile up instead of flowing over large extensive area as a result flow assumes irregular bulbous form with less [l/t] ratio because of its less lateral extent. The top surface of such flows will be irregular instead of being plain and tabular and will have slopes in all directions because of extensive elevations and depressions. The top surfaces of these flows will not be smooth but will show irregularities. The nature of bottom of flow is determined by nature of land surface on which it is outpoured. However, bottom of such flows also exhibits irregular pattern. If supply of lava flow is in less quantity then it will show limited lateral extent. Continual eruption of this type of flow will result in piling of small flows with irregular form with bottoms showing conspicuous undulations and irregularities.

In view of these considerations an attempt has been made to determine the nature of Deccan Trap volcanictiy on the basis of the field characters exhibited by lava flows in the four ghat sections of study area.

Lateral Extent and Length to Thickness [l/t] Ratio.

As can be been seen from detailed descriptions of basalt flows exposed in the four ghat sections of study area as given in chapter II there is a lot of variation in their thickness, lateral extent and l/t ratio. Most of the flows are found to be of Compact Basalt. These flows are thick and are extensive showing high l/t ratio. There are some few flows of Amygdaloidal Basalt which are observed and have limited lateral extents and low l/t ratio. It is also observed that these characters of amygdaloidal basalt are governed by gas cavities. However, the compact basalt flows are free from gas activities are thick and are extensive too. Vesicular basalt and amygdaloidal basalt have limited lateral extent with low l/t ratio. Thus variation in thickness, level extent and l/t ratio clearly indicate that different type of volcanic activities had given rise to different type of flows.

While considering the morphology of the flows of the four ghat sections of study area there is a significant difference in fluidity of lava which after consolidation resulted in the formation of different types of flows, commensurable with different type of volcanicity.

MORPHOLOGY OF BASALTS COMPACT BASALT FLOWS

1] Morphology of compact basalt flows:

The thick extensive flows of compact basalt in the four ghat sections of study area exhibit regular tabular form and their tops as a whole are plane surfaces. The flows in all four ghat sections have considerable lateral extent and are traceable for variable distances. All flows could not be traced laterally because of steep slopes. However, on the basis of topographical expressions they could be recognized from long distance as compared to lateral extent these flows have small thickness showing large l/t ratio.

The compact basalt flows occurring in all four ghat sections exhibit regular tabular form. The tabular form thus indicates cooling and consolidation of large quantity of fluid lava, Their extensive lateral spread, small thickness, high l/t ratio and absence of vesicularity except in top portion of individual lava flows are indications of the fact that compact basalts are products of fissure type of volcanicity characterized by outpouring of basaltic lavas in enormous quantity that spread over a few hundred square kilometres in thick tabular form.

2] Predominance of compact basalt flows:

Although there is predominance of compact basalt

flows in all four ghat sections of study area amygdaloidal basalt are also occurring in the form of bulbous and undulated flows. Not only Amygdaloidal basalt but bands of red tachylyte are also seen. Their occurrences are given in Table No.6.1 The thickness of compact basalt in the various ghat sections in the study area is mentioned below. Whereas amygdaloidal basalt flows of compound lava flows have comparatively very less thickness. Amygdaloidal flows are thin irregular flows.

Table No 5.1

Basalt Flow Thickness Present in Ghat Sections

Name of Ghat Section	**Total Thickness inmeters**	**Total Thickness of Compact Basalt Flow in meters**	**Total Thicknessof Amygdaloidal Basalt Flows in meters**
Vambhori/ Dongargan	106.5	104.37	2.13
Garmatha	135	103.50	31.00
Sautada	150	139.28	9.22
Imampur	118	100.00	16.40

3] Origin of Compact Basalt Flows:

Thick and extensive lava flows of compact basalt having high l/t ratio in all the ghat sections are indicative of fissure type of volcanicity. Thus floods of enormous quantities of basaltic lavas have spread over few hundred square kilometres resulting into the formation of thick tabular sheets of basalts. This clearly shows that the fissure type of volcanicity was highly predominant. However, there are two types of compact basalt flow which are predominantly seen in the ghat sections. They are

A] Compact Porphyritic Flow

B] Compact Aphanitic Flow

Compact porphyritic flows are predominantly seen in the Vambhori-Dongargan ghat section and Imampur ghat section respectively. They might be the product of early plagioclase fractionation and relatively slow cooling rate. The compact aphanitic flows are highly predominant in Sautada ghat section and Garmatha ghat section, which may be an indication of early olivine fractionation and rapid cooling of magma. The porphyritic flows in Imampur ghat section are lenticular shaped and few have small lateral extent which may be due to outpouring of lava in small quantity. This shows that in all the four ghat sections above M.S.L. of 600 m. Fissure type of volcanic activity was predominant.

AMYGDALOIDAL BASALT FLOWS

As compared to compact basalt amygdaloidal basalt flows show quite different morphology. Their undulated irregular pattern with limited lateral extent and low l/t ratio is an indication that they are the products of outpouring of viscous lava in less quantity and they are not the product of fissure type of eruption. The lava which is viscous in nature must have been outpoured through large number of outlets and are represented by vertical or inclined tail like intrusions connected with the base of some of these flows and cutting underlying flows. [Fig. No. 2.14] These type of volcanic activities were significant in the ghat section above M.S.L. 650 m up to M.S.L. 700 m.

In addition to the flows given in Table No. 5.1 there are intrusions of Red Tachylytic Basalt having thickness up to 30 cm. in the form of pockets and lenses and are described below in Table No. 5.2

Table No. 5.2

Red Tachylytic Basalt Flows in Ghat Section

Name of Ghat Section	Flow No. of Red Tachylyte
Imampur	2, 22, 29 and 31
Sautada	1 (Green Tachylyte) and 6
Garmatha	3

However, Vambhori/Dongargan ghat section is devoid of occurrence of Tachylyte.

Lateral Extent of Type of Volcanicity:

To determine lateral extent of the type of volcanicity in the area, the basalt flow occurring the same elevations have been take into considerations.

Imampur Ghat Section:

Imampur ghat section is located in the district of Ahmadagar in the state of Maharashtra, It is available on Aurangabad-Ahmadnagar highway. The ghat section is 87.700 km away from Aurangabad city towards Ahmadnagar and 24.300 km away from Ahmadnagar city while going towards Aurangabad. The ghat section starts appearing from village 'Pandhrichapool' which is situated at the base at the height M.S.L. 605 m The ghat section ends at village Imampur which in situated at top

M.S.L. 723 m. In all 32 lava flows are exposed in the ghat, there are 9 thick and thin lava flows of Amygdaloidal basalt the occurrence of Amygdaloidal basalt flows is limited and their average lateral extent is small up to 2.5 m only (Fig No. 2.14, Fig. No. 2.15) of Flow No. 26 and Flow No.27 the average thickness of amygdaloidal lava flows is 1.71 meters. However, it is evident that these amygdaloidal lava flows are the lava in small quantity (Fig No. 2.9, Flow No. 17; Fig. No. 2.14, Flow No. 26; Fig. No. 2.15 Flow No. 27 respectively). Amygdaloidal basalt makes 15.25% of the entire Imampur ghat section.

In the Imampur ghat section there are 15 flows of compact porphyritic basalt making 70.47% of entire ghat section. These flows have average thickness of 5.10 meters. However, the maximum thickness of 40 meters is seen. (Flow No. 1 Fig. No. 2.1) Few of them are lenticular in shape.

There are four flows of compact Aphanitic basalt in entire Imampur ghat section and these four flows making 12.28% of the entire ghat section.

Table No. 5.3

Flow Type, Thickness and Their Percentage in Imampur GhatSection

Flow Type	Total No. of Flow	Maximum Thickness in meter	Minimum Thickness in meter	Average Thick-ness in meter	% of Flow Type in Entire Ghat Section
Amygdaloid al Basalt	9	3 meter	0.5 meter	1.71 meter	15.25%
Compact Aphanitic Basalt	4	6 meter	2 meter	3.62 meter	12.28%
Red Tachylyte	4	0.5 meter	0.5 meter	0.5 meter	2.00%
Compact Porphyritic Basalt	15	40 meter	1 meter	5.10 meter	70.47%

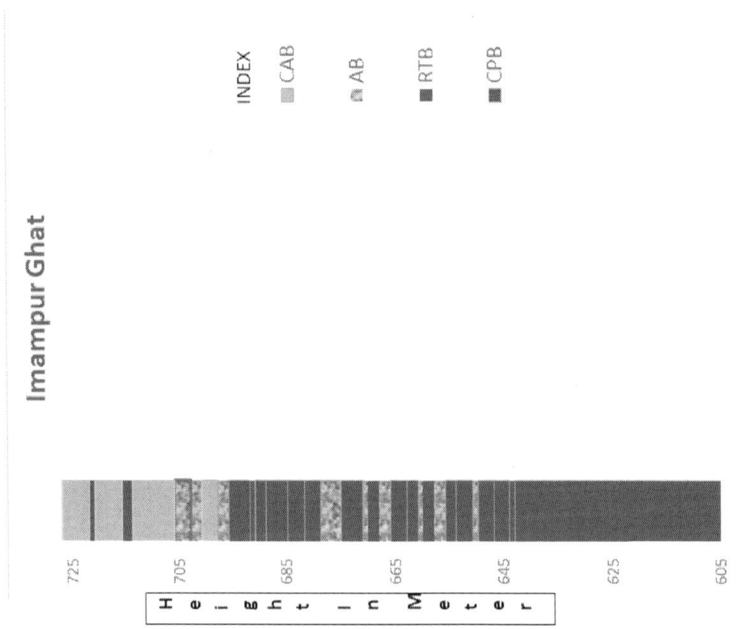

Fig. No. 6.1 Sequence of Lava Flows in Imampur Ghat

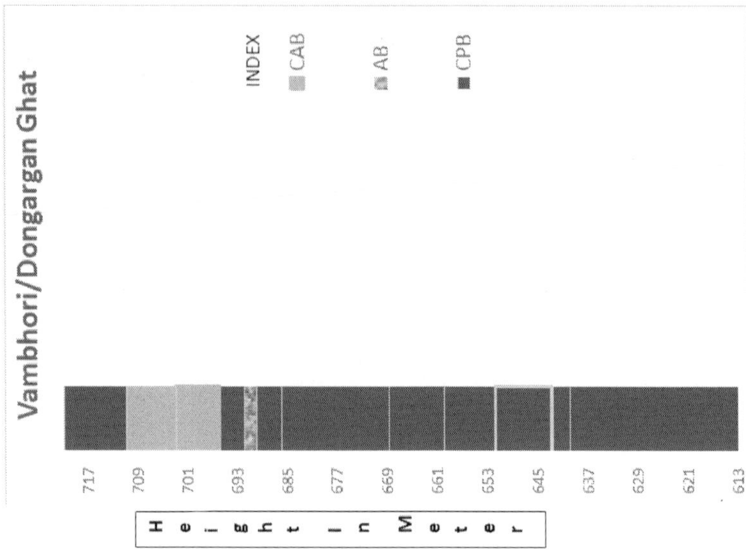

Fig. No. 6.2 Sequence of Lava Flows in Vambhori / Dongargan Ghat

VAMBHORI / DONGARGAN GHAT SECTION

Wamhori Dongargan ghat section is exposed on Ahmadnagar- Shendi – Dongargan - Vambhori road. It is 21.5 km away from Ahmadnagar. The ghat section starts from Vambhori village river bridge which is at base [ch. 21.500] M.S.L. 613 m to Dongargan village at top M.S.L. 719.50 m of [ch.19]

There are twelve lava flows exposed in the entire ghat section out of which 9 lava flows are of porphyritic basalt making 83.64% of entire ghat section. [Between M.S.L. 613 m to M.S.L. 719.50 m] There are nine porphyritic basalt lava flows present and these porphyritic basalt flows have average thickeness is of 9.89 m. All the porphyritic basalt flows have well developed plagioclase phenorysts ranging from 7.5 cm to 10 cm in length.

There are two Aphanitic lava flows exposed along the ghat section. Maximum thickness exhibited is of 8.30 m. and average thickness is 7.65 m. These two Aphantic basalt flows makes 14.36% of entire ghat section.

There is only one flow of Amygdaloidal basalt exposed in the section having thickness of 2.82 m. making 2.00% of entire ghat section. Absence of Tachylytic basalt flow, presence of plagioclase as giant phenocryst are the features of the lava flows in Vambhori- Dongargan ghat section.

Table No. 6.4

Flow Type, Thickness and Their Percentage inVambhori/ Dongargan Ghat Section

Type Flows	No. of Flows	Flow With Max. Thickness	Minimum Thickness	Average Thickness	Flow% in Ghat Section
Amygdaloidal Basalt	1	2.82 meter	2.82 meter	2.82 meter	2.64%
Compact Aphanitic Basalt	2	8.30 meter	7.00 meter	7.65 meter	14.36%
Compact Porphyritic Basalt	9	27.00 meter	2.75 meter	9.82 meter	83.00%

SAUTADA GHAT

Sautada ghat section is exposed on Ahmadnagar-Jamkhed- Beed

The ghat section is available in outskirts of Jamkhed at a distance of 76 km. from the township of Ahmadnagar. The ghat section starts appearing at M.S.L. 627.43 m at its base and its top demarcates the boundary between Ahmadnagar district and Beed district of Marathwada region M.S.L. 777.43 m.

There are 10 lava flows exposed along the ghat section, out of which 6 flows are compact Aphanitic basalt flows making 92.52% of entire ghat section and the average thickness of compact Aphanitic basalt flow is 23.33 m.

There are only two Amygdaloidal Basalt lava flows in this section. The maximum thickness of the Amygdaloidal lava flow is 8.22 m and these flows makes 6.82% of entire ghat section. Entire ghat section is devoid of any porphyritic basalt flow. Summary of the flows is given in Table No.5.5

Table No. 5.5

Flow Type, Thickness and Their Percentage in Sautada GhatSection

Flow Type	No. of Flow	Max. Thickness of Flow	Minimum Thickness Flow	Average Thickness	Flow% in Ghat Section.
Amygdaloidal Basalt	2	8.22 meter	2 meter	5.11	6.82%
Red Tachylyte	2	0.57 meter	0.5 meter	0.53 meter	0.66%
Compact Aphanatic Basalt	6	49.70 meter	6.10 meter	23.33 meter	92.52%

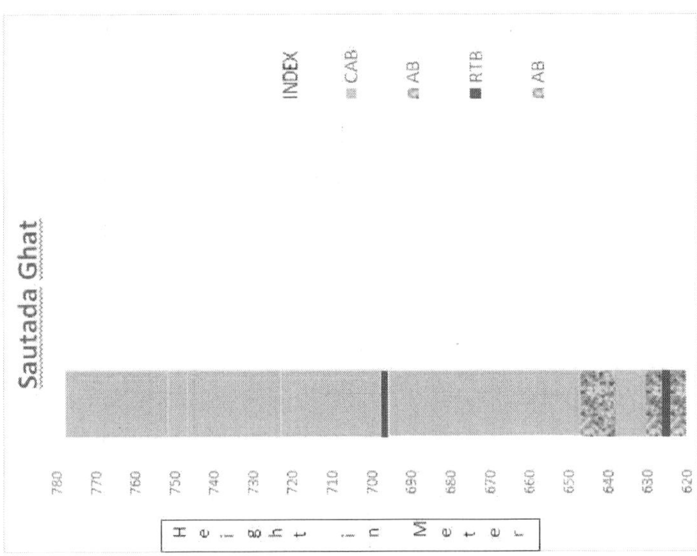

Fig. No. 6.3 Sequence of Lava Flows in Sautada Ghat

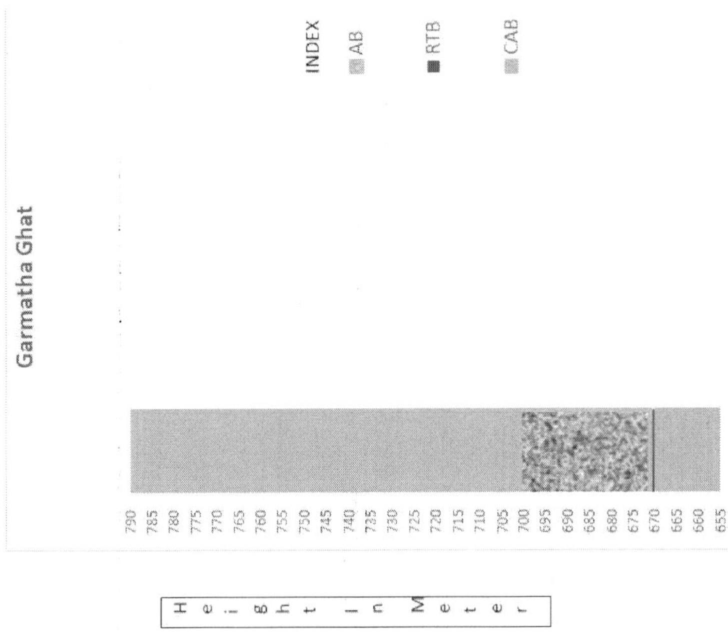

Fig. No. 6.4 Sequence of Lava Flows in Garmatha Ghat

GARMATHA GHAT

This ghat section is exposed on Beed-Ahmadangar road in the district of Beed in Marathwada region. The ghat section is available at a distance of 50 km. from township of Beed at village Dongarkinhi at base M.S.L. 655 and ends at village Bhatepura situated at top of ghat section at M.S.L. 790 m.

There are 6 lava flows exposed along the ghat section most of them have large lateral exent and flat surfaces. Out of 6 lava flows 4 lava flows are compact aphanitic basalt flows with maximum thickness of is 55 m. and average thickness of 25.88 m. Compact Aphanitic basalt flow makes 76.25% of entire ghat section. There is only one flow of Amygdaloidal basalt having thickness of 31 m and making 23% entire ghat section. Entire ghat section is devoid of porphyritic basalt flow. The summary of the lava flows is given in Table No.5.6.

Table No. 5.6

Flow Type, Thickness and Their Percentage in Garmatha GhatSection

Flow Type	No. of Flow	Max. Thicknessof Flow inmeter	Minimum Thickness of Flow in meter	Average Thicknessof Flow inmeter	Flow % in Ghat Section in meter
AmygdaloidalBasalt	1	31 meter	31 meter	31 meter	23.00%
Red Tachylyte	1	0.5 meter	0.5 meter	0.5 meter	0.75%
Compact Apharitic Basalt	4	55 meter	6 meter	25.88 meter	76.25%

Comparison With Chikhaldara and Toranmal Ghat Sections

To determine the lateral extent and type of volcanicity that prevailed in the ghat sections of the study area, the Basalt flows occurring at the same elevations in the two neighbouring areas i.e. Toranmal ghat section (A.V.Tejankar 2002) and Chikhldara ghat section (S.D.Doke 2013) have been taken into considerations.

Toranmal Ghat section is available in Nandurbar district present in North West part of Maharashtra state. It is Located at a distance of 372.5 km. from Imampur ghat section. In Toranmal ghat section the stretch i.e from M.S.L. 600 m to M.S.L. 780 m thick extensive flows of Compact Basalt are making up to 100% [90% Compact Porphyritic Basalt and 10% Compact Aphanitic Basalt respectively]

Chikhaldara ghat section is available in Amravati district present in North East part of Maharashtra state. It is located at a distance of 480km. from Sautada ghat section. In this ghat section for the same stretch Compact Aphanitic Basalt makes 6.66%, Amygdaloidal Basalt makes 16.66%, and Compact Porphyritic Basalt makes 76.68% respectively (Fig. No. 5.1)

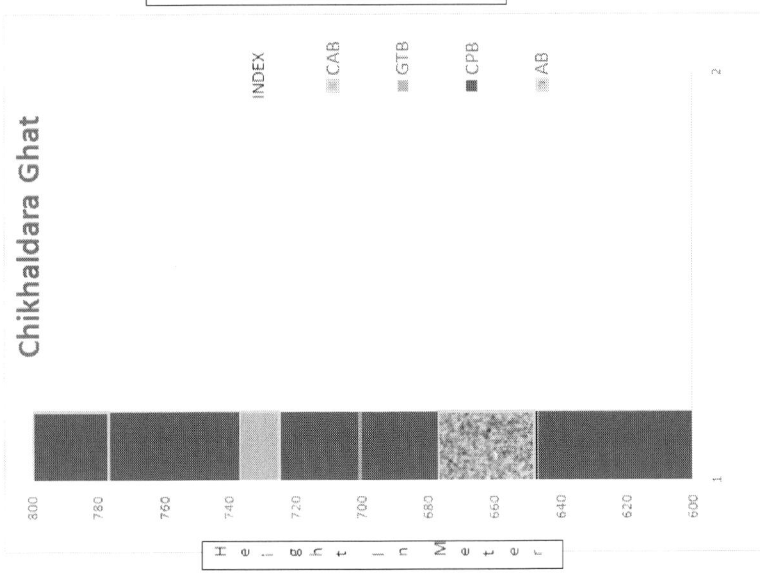

Fig. No. 6.5 Sequence of Lava Flows in Chikhaldara Ghat

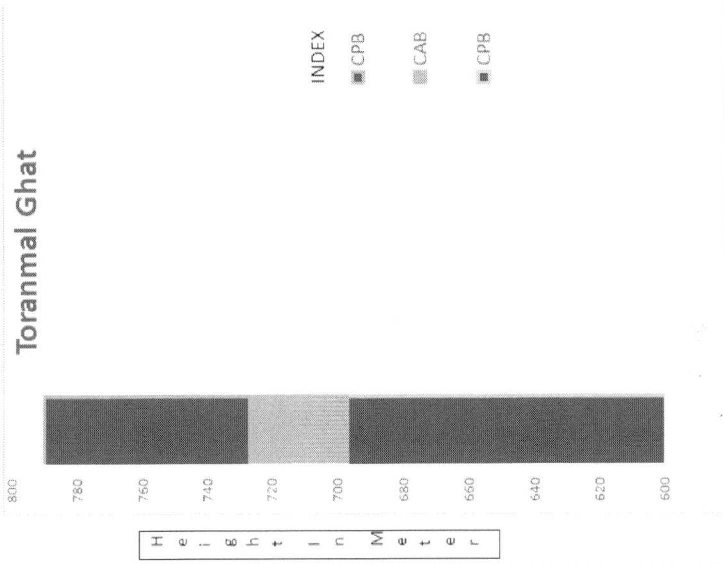

Fig. No. 6.6 Sequence of Lava Flows in Toranmal Ghat

For the same elevation comparison of lava flow type and their percentage in between these two ghat sections is given below in Table No.5.7

Table No. 5.7
Comparative Study of Lava Flows in Toranmal and ChikhaldaraGhat Section

Flow Type	Flow % in Toranmal Ghat Section	Flow % in Chikhaldara Ghat Section
Compact Porphyritic Basalt	90%	67.68%
Compact Aphanitic Basalt	10%	6.66%
Amygdaloidal Basalt	Nil	16.66%

Comparison with Manjarsumba and Nagad –Nagapur ghat sections.

To know the nature of volcanicity occurred in Beed district of Marathwada region basalt flows occurring at same elevation of Manjarsumba ghat section have been taken into consideration. (P.S. Kulkarni 1984)

Manjarsumba ghat section is available on Beed-Osmanabad road when one is facing towards Osmanabad. It is 12.500 Km. away from the township of Beed. Though the ghat section is available from M.S.L. 576 m. with four flows, for comparative account flow above M.S.L. 608 m have been taken into consideration. In this ghat section from M.S.L. 608 m to M.S.L. 717 m three flows are available. The flow at base is compact aphanitic basalt flow having thickness of 38 m from M.S.L. 608 m to M.S.L. 646 m and it is making 34.86% of the ghat section. The topmost flow is compact porphyritic basalt flow having thickness of 70.50 m making 64.67% of the ghat section. Both the flows have been separated by Red Tachylytic basalt flow having thickness of 0.5 m at M.S.L. 646 m to M.S.L. 646.5 m making 0.44% of the ghat section.

Nagad-Nagapur ghat section is available on Chalisgaon – Nagad - Nagapur road in Aurngabad district of Marathwada region. (Approach Road to Nagad – Nagapur is from Chalisgaon town of Jalgaon District of Khandesh Region) It is 90 Km. away from the township of Aurangabad towards its North. Though the ghat section is available from M.S.L. 390 m with eighteen flows for comparative account the flow above

M.S.L. 605 m is taken for consideration. In this ghat section from M.S.L. 605 m to M.S.L. 668 m there are four thick and thin irregular Amygdaloidal basalt flows making 60% the ghat section. From M.S.L. 668 m to M.S.L. 670 m there is a thin band of Red Tachylyte making 2% of the ghat section. From M.S.L.670 m to 710 m there is one thick flow of compact Porphyritic basalt having thickness of 40 m with medium grained plagioclase phenocryst. The flow makes 38% of the ghat section.

For the same elevation comparison of lava flow type and their percentage in between these two ghat sections is given below in Table No. 5.8

Table No. 5.8
Comparative Study of Lava Flows in Manjarsumba and Nagad-Nagapur Ghat Section

Flow Type	Flow % in Manjarsumba GhatSection	Flow % in Nagad – Nagapur Ghat Section
Compact PorphyriticBasalt	64.67%	38.00%
Compact AphaniticBasalt	34.86%	Absent
Amygdaloidal Basalt	Absent	60.00%
Red Tachylyte	0.44%	2.00%

Vertical variation in nature of volcanictiy in Deccan Trap area

As can be seen from above all observations, thick extensive flows of Compact Basalt representing fissure type of volcanic activity are occurring above M.S.L. 600 m This is in agreement with central part of Deccan Trap (Mhaismal, Kulkarni P.S. 1984) North West part of Deccan Trap (Toranmal, Tejankar A.V. 2002) and North East part of Deccan Trap (Chikhaldara, Doke S.D. 2013)

It is interesting to see that in all the ghat sections of these areas of Deccan Trap the flows of Compact basalt

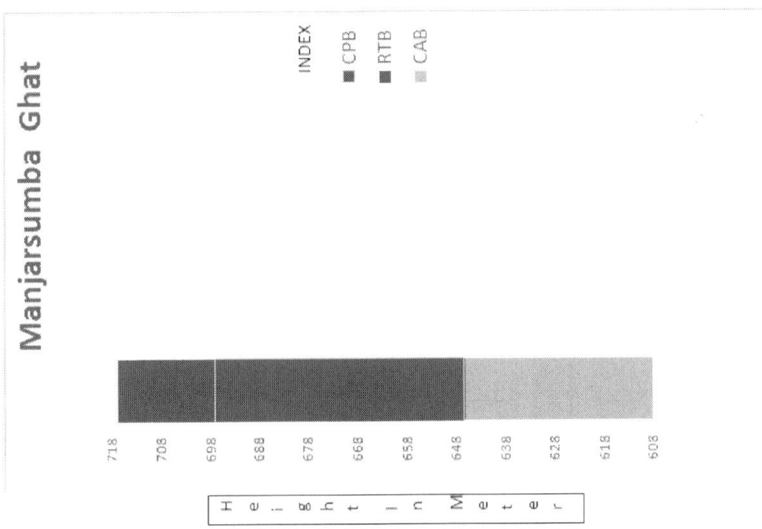

Fig. No. 6.7 Sequence of Lava Flows in Manjarsumba Ghat

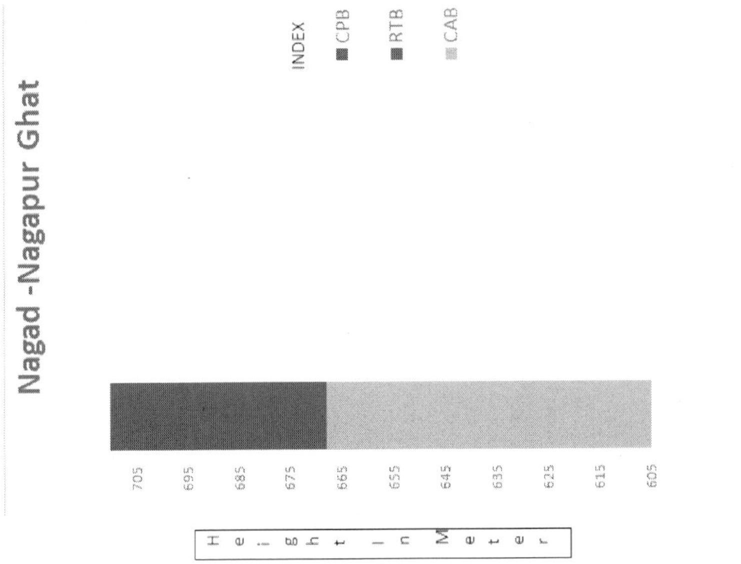

Fig. No. 6.8 Sequence of Lava Flows in Nagad- Nagapur Ghat

representing fissure type of volcanic activity are predominantly occurring at higher levels i.e. above M.S.L. 600 m.

VARIATION IN THE NATURE OF VOLCANICITY IN DECCAN TRAP AREA

1. It has been observed that thick exclusive compact Basalt flows are predominating in area under study. Fissure type of volcanic activity seems to be prevalent occurring above M.S.L. 600 up to M.S.L. 790 m. This is in agreement with central part of Deccan Trap (P.S. Kulkarni 1984)

2. As stated earlier though the distance in between the ghat sections is in the range of 10 km to 90 km. in the study area, despite of predominance of Compact Basalt, they exhibit different field characters and lithology. The compact basalt flows are not extensive as they were supposed to be which means that different areas (ghat sections) were served with different feeder system. (Marathe 1975)

3. The occurrence of few lenticular shaped porphyritic Basalt flows in Imampur ghat section may indicate the outpouring of lava in small quantity without gases and volatile matter in to it.

4. The predominance of lath shaped Plagioclase phenocrysts in Vambhori-Dongargan ghat section

and Imampur ghat section may be an indication of relatively slow rate of cooling of lava flows as compare to lava flows in Sautada and Garmatha ghat sections respectively.

5. The Predominance of lath shaped and large sized Plagioclase phenocrysts in the lava flows of Vambhori/ Dongargan ghat section may indicate the accumulation of plagioclase phenocrysts by floatation in highly differentiated (denser than plagioclase) magmas that concentrated near the roofs of magma chambers at shallow depth which is in agreement with R. K. Sharma and Sudha Vaddadi. (1996)

6. Zoning, and melt inclusions in the plagioclase phenocrysts of Imampur and Vambhori/ Dongargan Ghat Sections may indicate that dissolution may have occurred during the formation of plagioclase phenocrysts due to their longer time of residence in the magma chamber as compared to the lava flows in Sautada and Garmatha Ghat sections may indicate different sources of magma for the four ghat sections of the study area. (Marathe 1975)

7. The basalt, in general, is having more or less Tholeiite composition. The differentiation trend is characterised by moderate Mgo enrichment. The basalts are characteristically low in K_2O and in this respect are similar to OIT and MORB

Tholeiites. These flows also show similarities with the basalts of Salbardi, Mahabaleshwar, Sagar and Hyderabad in Eastern, South eastern and Western parts Trappean regime respectively. Which is an agreement with Wilson (1989).

8. The lava flows here as a whole are of derivative in nature possibly generated from fractionation of Mgo–rich parental magma derived from mantle source. Variation in major element chemistry with wide variation in some incompatible elements may suggest fractionation in a near surface magma chamber.

9. The trace element viz, Ba, Rb and Sr exhibit fairly wide variation in comparison to the limited range of variation observed in major elements. This feature may be an indicative of open system fractional crystallization process in a near surface magma chamber

fed regularly with the batches of parental magma which has mingled with the liquids already present. Which is in agreement with L. S. Shekhavat and V. P. Sharma 1996.

10. The presence of Spherulites only in aphanitic basalt lava flows of Imampur and Garmatha ghat sections could be an indication of devitrification of brown basaltic glass during the process of cooling of lava.

11. There is a general incidence of flow with average Mg. No. ranging in between 44.42 and 49.10 may indicate early Olivine fractionation followed by opx, cpx fractionation and plagioclase fractionation respectively which is in agreement with Cox, 1980.

12. The ratio of various trace elements also confirms the same order of magmatic fractionation process.

13. The evidence for contamination, possibly by crustal material is noted. The lava flows here in general exhibit CaO/Al_2O_3 values ranging from 0.55 to 0.77 may indicate varying degree of contamination which is in agreement with S.S. Deshmukh, T. Sano and K.K.K. Nair, 1996.

14. Even though over large areas for longer periods fissure eruptions have occurred, the sequence of the flow in each ghat section is different.

15. The compact basalt flows are not extensive as they were supposed to be which means that different areas (ghat sections) were served with different feeder system. (Marathe 1975)

16. Presence of Red Tachylyte without baked soil or tuffs in the ghat sections of study area is in agreement with the conclusion drawn by Kulkarni (1975) that the lava flows came in quick succession without Intertrappean intervals.

17. The field characters of basalt flows of the ghat

sections of study area have clean basalt to basalt contact at many places which confirms the view that the lava flows came in quick succession and there were no intertrappean intervals as concluded by P.S. Kulkarni. (1984)

18. The melting range of magma may have been 60 Km. to 100 Km. below the surface.

BIBLIOGRAPHY

Al Marhoun, M.A.: *Pressure-volume-temperature Correlations for Saudi Crude Oils*, Society of Petroleum Engineers Publication, UK, 2005.

Anderson, Mary P. and William W. Woessner: *Applied Groundwater Modeling: Simulation of Flow and Advective Transport.* Academic Press, San Diego, 1992.

Arnold, C. A.: *Introduction to Paleobotany*, New York, McGraw-Hill, 1947.

Asaithambi S.: *Economics of Ground Water Management in India*, Abhijeet, Delhi, 2008.

Babar Md. : *Hydrogeomorphology : Fundamentals Applications and Techniques*, New India Pub, Delhi, 2005.

Baker, Victor R. : *Catastrophic Flooding: The Origin of the Channeled Scabland*, Stroudsburg, Pa., Dowden, 1981.

Baruah Akhil : *Advanced Morphology of Angiosperms*, Aavishkar, Delhi, 2008.

Berkoff, Jeremy: *A Strategy for Managing Water in the Middle East and North Africa*, World Bank, Washington, DC, 1994.

Best, Myron G.: *Igneous and Metamorphic Petrology*, Blackwell Publishing, Delhi, 2002.

Bhakar, S.R. : *Ground Water Hydrology : Theory and Practice*, Agrotech, Delhi, 2009.

Bhave, P.R. and R. Gupta: *Analysis of Water Distribution Networks*, Narosa, Delhi, 2011.

Biswas Asit K. : *Integrated Water Resources Management in South and South-East Asia*, , Oxford University Press, Delhi, 2001.

Blatt, Harvey; Tracy, Robert J.; Owens, Brent: *Petrology: Igneous,*

Bouma, A. H.: *Sedimentology of Some Flysch Deposits*, Elsevier, Amsterdam, 1962.

Bouma, A., Normark, W. R., and Barnes, N. E.: *Submarine Fans and Related Turbidite Systems*, Springer Verlag, New York, 2006.

Bourne, Peter G.: *Water and Sanitation: Economic and Sociological Perspectives* Academic Press. Orlando. 1984.

Brantly, John Edward: *History of Oil Well Drilling*, Book Division, Gulf Pub. Co., Houston, 2011.

Brian, Moss: *Ecology of Fresh Waters – Man and Medium*, Past to Future, NY, 1997.

Brooks, D.: *Water: Local-Level Management*, Ottawa, International Development Research Centre, 2002.

Bull, William B., and McFadden, Leslie D.: *Tectonic Geomorphology of the Garlock Fault*, Publications in Geomorphology, New York, 2007.

Burdick, Donald L., and William L. Leffler: *Petrochemicals in Nontechnical Language,* PennWell, Tulsa, Okla, 2011.

Carter, David : *Dynamic Symmetry,* Exhibition Catalog, 1961.

Chappell, J. : *Geology of Coral Terraces, Huon Peninsula, New Guinea; A Study of Quaternary Tectonic Movements and Sea-level Changes*, Geological Society of America Bulletin, 1974.

Clark, James Anthony, C. A. Warner, and H. E. Walton: *The Chronological History of the Petroleum and Natural Gas Industries,* Clark Book Co., Houston, 2003.

Cloud, P. : *Oasis in Space: Earth History from the Beginning,* New York, Norton Books, 1988.

Conaway, Charles F.: *The Petroleum Industry: A Nontechnical Guide,* PennWell Pub. Co., Tulsa, OK, 2009.

Conway, D.: *Climate Change and Water Resources in the Nile Basin*, University of London, London, 1993.

Conway, D.: *Climate Change and Water Resources in the Nile Basin*, University of London, London, 1993.

Cordesman, Anthony H.: *The Gulf and the West: Strategic Relations and Military Realities*, Westview Press, Boulder, Colo., 2008.

Cullis, A. : *Rainwater Harvesting: The Collection of Rainfall and Runoff in Rural Areas,* London, U.K.: IT Publications, 1986.

Dabrowski, Magdalena : *Contrasts of Form: Geometric Abstract Art 1910- 1980*, New York: Museum of Modern Art, 1985.

Datta-Barua, Lohit: *Natural Gas Measurement and Control: A Guide for Operators and Engineers*, McGraw-Hill Professional Book Group, London, 2011.

Davies R. J., Cartwright J., Stewart S. A, Underhill, J. R., Lappin M.: *3D Seismic Technology: Application to the Exploration of Sedimentary Basins*, Geological Society, London, Memoir, 2013.

Davis, John: *Oil and Canada-United States Relations*, National Planning Association, New York, 2009.

Devereux, Steve: *Drilling for Oil & Gas: A Nontechnical Guide,* PennWell, Tulsa, OK, 2009.

Diener, Daniel: *Introduction to Well Control,* Petroleum Extension Service, University of Texas at Austin, Austin, 2009.

Diessel, C.F.K.: *Coal-bearing Depositional Systems.* Springer-Verlag, Berlin, 2012.

Doneen L.D.: *Water Quality for Agriculture*, Department of Irrigation, University of California, Davis, 1964.

Donnez, Pierre: *Essentials of Reservoir Engineering,* Editions Technip, Paris, 2017.

Dunai, T.J.: *Cosmogenic Nucleides,* Cambridge University Press, U.K., 2016.

Dunai, T.J.: *Cosmogenic Nucleides,* Cambridge University Press, U.K., 2010.

Dunn, J.A. and A.K. Dey: *The Geology and Petrology of Eastern Singhbhum and Surrounding Areas*, Kolkata, 2001.

Dunne, T. and L.B. Leopold: *Water in Environmental Planning*, W.H. Freeman and Co., San Francisco, CA, 2008.

Dunne, T. and L.B. Leopold: *Water in Environmental Planning*, W.H. Freeman and Co., San Francisco, CA, 1978.

Dunning, William V. : *Changing Images of Pictorial Space:* Syracuse, Syracuse U. Press, 1991.

Dyer, M. I., and D. A. Crossley: *Coupling of Ecological*

Studies with Remote Sensing: Potentials at Four Biosphere Reserves in the United States, Washington, 1986.

Eckenfelder W.W: *Principles of Water Quality Management*, CBI Pub. Co., Massachusetts, 1980.

Edmaier, Bernhard: *Earthsong*, Phaidon Press, London, 2014.

Edwards, Denis : *Earth Revealing Earth Healing: Ecology and Christian Theology,* Collegeville: Liturgical Press, 2001.

Edwards, Denis: *Earth Revealing Earth Healing: Ecology and Christian Theology,* Collegeville: Liturgical Press, 2011.

Edwards, Edward : *Pattern and Design with Dynamic Symmetry,* NY: Dover 1967.

Egan, Walter G.: *Photometry and Polarization in Remote Sensing*, New York, 2005.

Eisner, A.: *Hydrogenation and Petrography of Some Low-rank Coals from the Western United States,* Bureau of Mines, Washington, D.C., 2010.

Engel, M.H., and S. Macko: *Organic Geochemistry; Principles and Applications*, Plenum Press, New York, 2013.

Erik Nissen-Peterson : *Rainwater Catchment Systems,* UK: Intermediate Technology Publications, 1999.

Ezekwe, Nnaemeka: *Petroleum Reservoir Engineering Practice,* Prentice Hall, Upper Saddle River, NJ, 2016.

Feder, Paul A.: *A Guidebook to Groundwater: Resources and Education Opportunities in the Great Lakes Region*, Great Lakes Commission, Ann Arbor, 1993.

Fenichell, Stephen: *Plastic: The Making of a Synthetic Century,* Harper Business, New York, 2006.

Fetter, Charles Willard: *Applied Hydrogeology*, Merrill Publishing Company, Columbus, OH. 1988.

Ford, Trevor D. and Cullingford, C. H. D. (editors), *The Science of Speleology:* Academic Press, New York, 593 p.

Frasier, Gary, and Lloyd Myers: *Handbook of Water Harvesting*, U.S. Dept. of Agriculture, Agricultural Research Service, Washington D.C., 1983

Frasier, Gary, and Lloyd Myers: *Handbook of Water Harvesting*, U.S. Dept. of Agriculture, Agricultural Research Service, Washington D.C., 1983

Friis, E.M. and P.K. Endress : *Origin and Evolution of Angiosperm Flowers*, Advances in Botanical Research, 1990.

Gautam, Ashutosh: *Water and Wastewater Analysis*, Daya, Delhi, 2002.

Geraghty, James J.: *Water Atlas of the United States*, Port Washington, NY, 1973.

Ghyka, Matila : *The Geometry of Art and Life,* NY: Dover, 1977.

Gischler, Christiaan E.: *Water Resources in the Arab Middle East and North Africa*, Cambridge, Eng, Middle East & North African Studies Press, 1979.

Glenn T. Trewartha : *Elements of Geography*, New York, McGraw Hill, 1941.

Goldsmith, Mike : *Stars and Planets,* New York: Kingfisher, 2008.

Golledge, R. and R. Stimson: *Spatial Behavior: A Geographic Perspective.* New York: The Guilford Press, 1997.

Gould, J., and E. Nissen-Petersen : *Rainwater Catchment Systems for Domestic Supply: Design, Construction and Implementation,* London, U.K.: IT Publications, 1999.

Gray, Forest: *Petroleum Production in Nontechnical Language,* PennWell Books, Tulsa, OK, 2015.

Guo, Boyun, William C. Lyons, and Ali Ghalambor: *Petroleum Production Engineering: A Computer-assisted Approach,* Gulf Professional Pub., Burlington, MA, 2016.

Gupta K.R.: *Water Crisis in India*, Atlantic Pub, Delhi, 2008.

Gupta, Alok K. : *Petrology and Genesis of Igneous Rocks*, Narosa, Delhi, 2007.

Gurjar, Ram Kumar : *Geography of Water Resources*, Rawat, Delhi, 2008.

Gurney, R. J. : *Remote Sensing in Hydrology*, New York, Van Nostrand Reinhold, 1991.

Hall, Edward T. : *Spatial Analysis in Archaeology,* Cambridge: Cambridge University Press, 1976.

Hallam, A.: *A Revolution in the Earth Sciences:* New York, Oxford University Press, 2013.

Hallam, A.: *A Revolution in the Earth Sciences:* New York, Oxford University Press, 1973.

Hambidge, Jay : *The Elements of Dynamic Symmetry,* NY: Dover, 1967.

Hanks, Kurt : *Rapid Viz: A New Method for the Rapid Visualization of Ideas,* CA: Kaufmann, 1980.

Hannum, M.B.: *Late Quaternary Evolution of the Kennebec and Damariscotta River Estuaries*, Geological Sciences, Univ. Maine, Orono, 2007.

Hannum, M.B.: *Late Quaternary Evolution of the Kennebec and Damariscotta River Estuaries*, Geological Sciences, Univ. Maine, Orono, 1997.

Hansen D.M., Cartwright, J.A, and Thomas, D. 2004. The development of intersecting igneous sills from 3D seismic. Geological Society of London, Special Publication. *In:*

Harrelson, C.C., Rawlins, C., Potyondy, J.: *Stream*

Channel Reference Sites: An Illustrated Guide to Field Technique, USDA Forest Service, General Technical Report RM-245, Fort Collins, Co., 1994,

Hartshorne, Richard: *Perspective on the Nature of Geography*, Chicago, Rand McNally, 2009.

Hartshorne, Richard: *Perspective on the Nature of Geography*, Chicago, Rand McNally, 1959.

Helen Bannayan: *Water Resources of Jordan: Present Status and Future Potentials*, Amman, Friedrich-Ebert-Stiftung, 1993.

Hemenway, Toby : *Gaia's Garden: A Guide to Home-Scale Permaculture*, Vermont: Chelsea Green Publishing Company, 2000.

Henley, E. J. and J. D. Seader: *Equilibrium Stage Operations in Chemical Engineering,* John Wiley and Sons, New York, 2011.

Hill, Carol A. and Forti, Paolo: *Cave Minerals of the World:* National Speleological Society, Huntsville, Alabama, 1986.

Hillel, Daniel: *Rivers of Eden: the Struggle for Water and the Quest for Peace in the Middle East*, Oxford University Press, NYC, 1994.

Hines, A. L., and R. M. Maddox: *Mass Transfer: Fundamentals and Applications*, Prentice-Hall PTR, Upper Saddle River, NJ, 2015.

Holden, Joseph: *Introduction to Physical Geography and the Environment*, Prentice-Hall, London, 2014.

Holden, Joseph: *Introduction to Physical Geography and the Environment*, Prentice-Hall, London, 2004.

Holmgren, D. A., Moody, J. D. : *The Structural Setting for Giant oil and Gas Fields*, Tokyo, 1975.

Huggett Richard John : *Fundamentals of Geomorphology*, T and F / Routledge, Delhi, 2008.

Hunt, J.M.: *Petroleum Geochemistry and Geology*, W.H. Freeman and Company, New York, 2006.

Isaac, J.: *Water and Peace in the Middle East: Proceedings of the First Israeli-Palestinian International Conference on Water, Zurich, Switzerland,* Elsevier, Amsterdam, 1994.

Issar, Arie: *Water Shall Flow from the Rock: Hydrogeology and Climate in the Lands of the Bible*, Springer-Verlag, NYC, 1990.

Jat M.L. and Bhakar S.R.: *Ground Water Hydrology: Theory and Practice*, Agrotech, Delhi, 2009.

Jat M.L. and Bhakar S.R.: *Ground Water Hydrology: Theory and Practice*, Agrotech, Delhi, 2009.

Jat, M.L. and S.R. Bhakar: *Ground Water Hydrology : Theory and Practice*, Agrotech, Delhi, 2009.

Johnsen, Ole: *Minerals of the World,* Princeton University Press, Princeton, NJ, 2000.

Juppenlatz, Morris : *Geographic Information Systems and Remote Sensing*, McGraw Hill, New York, 1996.

Kalwar S.C., Gurjar R.D. : *Geomorphology and Environmental Sustainability*, Concept, Delhi, 2005.

Kanmony, J. Cyril : *Drinking Water Management : Problems and Prospects*, Mittal Pub, Delhi, 2010.

Kappraff, Jay : *Connections: The Geometric Bridge between Art and Science,* NY: McGraw, 1990.

Kaur, Kanwaljit: *Handbook of Water and Wastewater Analysis*, Atlantic, Delhi, 2007.

Khilyuk, Leonid F., George V. Chilingar, and Herman H. Rieke: *Probability in Petroleum and Environmental Engineering*, Gulf Pub. Co., Houston, 2015.

Kondolf, G. Mathias; Hervé Piégay: *Tools in Fluvial Geomorphology*, Wiley, New York, 2003.

Kondolf, G. Mathias; Hervé Piégay: *Tools in Fluvial Geomorphology*, Wiley, New York, 2013.

Kuhn, Thomas S.: *The Structure of Scientific Revolutions,* Chicago: University of Chicago Press, 1970.

Kumar Yadav, Surendra : *Soil Ecology : Remote Sensing and Geographic Information System in Management,* APH, Delhi, 2007.

Langran, G.: *Time in Geographic Information Systems,* Taylor & Francis, Bristol, 2012.

Laurence M. Harwood, Christopher J. Moody: *Experimental Organic Chemistry: Principles and Practice*, WileyBlackwell, Oxford, 2009.

Leopold, L.B., M.G. Wolman, and J.P. Miller: *Fluvial Processes in Geomorphology*, Freeman, San Francisco, CA, 2004.

LeRoy, K. : *Sustaining Water Washington*, D.C.: Population Services International, New York, 1993.

Lillesand, T. M. and Kiefer, R. W.: *Remote Sensing and Image Interpretation*, New York, 1987.

Liverman, Diana L.: *People and Pixels: Linking Remote Sensing and Social Science.* National Academies Press, Delhi, 1998.

Lloyd Myers : *Handbook of Water Harvesting*, Washington D.C.: U.S. Dept. of Agriculture, Agricultural Research Service, 1983

Lunine, J. I. :*In Annual Reviews of Astronomy and Astrophysics*, California: Annual Reviews, 2002.

Lyell, C. : *Principles of Geology*, Boston, Little, Brown, 1853.

MacGillarvy, Caroline : *Symmetry Aspects of M.C. Escher's Periodic Drawings,* Utrecht: Uitgeversmaatschappij NV, 1965.

Mack, Pamela Etter: *Viewing the Earth: The Social Construction of the Landsat Satellite System*, Cambridge, MIT Press, 1900.

Mandelbrot, Benoit : *The Fractal Geometry of Nature,* San Francisco: Freeman, 1982.

Manthorp, P.A.: *Estuarine Circulation and Sediment Transport in the Saco River estuary*, Boston University, Boston, MA, 2005.

Manthorp, P.A.: *Estuarine Circulation and Sediment Transport in the Saco River estuary*, Boston University, Boston, MA, 1995.

Mathur, S M : *Concise Glossary of Geology*, Scientific, Delhi, 2007.

McGarry, M.G.: *Matching Water Supply Technology to the Needs and Resources of Developing Countries,* United Nations. New York. 1987

Meenu Bhatnagar: *Groundwater Management : Sustainable Approaches*, Icfai Books, Delhi, 2012.

Michael W. McElhinny: *The Earth's Magnetic Field,* New York: Academic Press, 2003.

Michael W. McElhinny: *The Earth's Magnetic Field,* New York: Academic Press, 1983.

Middlebrook, E.J., 1979. Industrial Water Pollution Control; J. Wiley & Sons, New York, U.S.A.

Miller, H. and S.-L. Shaw: *Geographic Information Systems for Transportation: Principals and Applications.* New York: Oxford, University Press.

Mitchum, R.M., Jr., Vail, P.R., and Sangree, J.B.:

Stratigraphy – Applications to Hydrocarbon Exploration, American Association of Petroleum Geologists, Memoir, 2007.

Morris, Mary E.: *Water Scarcity and Security concerns in the Middle East*, Abu Dhabi, Emirates Center for Strategic Studies and Research, 1998.

Naff T. and Matsow R.: *Water in the Middle East: Conflict or Cooperation*, Boulder, 1990.

Naiman, Robert J.: *The Freshwater Imperative: A Research Agenda*. Washington, D.C.: Island Press, 1995.

Narwani, G.S. : *Community Water Management*, Rawat, Delhi, 2005.

Nielsen, David M.: *Practical Handbook of Groundwater Monitoring*. Lewis Publishers, Chelsea, 1991.

Okoshi, T. : *Three-dimensional Imaging Techniques*, Academic Press Inc., New York, 1976.

Oldroyd, David : *Earth Cycles: A Historical Perspective*, Westport, Connicticut: Greenwood Press, 2006.

Palanisami, K : *Groundwater Management and Policies*, Macmillan Publishers India, Delhi, 2008.

Parikshit Ballabh: *Water Resource Management : Planning and Development*, Cyber Tech, Delhi, 2009.

Peters, K.E., and J.M. Moldowan: *The Biomarker Guide: Interpreting Molecular Fossils in Petroleum and*

Ancient Sediments, Prentice-Hall, Inc., Englewood Cliffs, 2003.

Peters, K.E., C.C. Walters, and J.M. Moldowan: *The Biomarker Guide,* Cambridge University Press, New York, 2015.

Peuquet, D.: *Representations of Space and Time*. New York: Guilford Press, 2012.

Philpotts, Anthony; Ague, Jay: *Principles of Igneous and Metamorphic Petrology,* Cambridge University Press, Cambridge, 2009.

Postnikov, M. : *Lectures in Geometry Semester I Analytic Geometry*, Moscow: MIR publishers, 1982.

Prasad, C.V.R.K.: *Elementary Exercises in Geology*, Universities Press, Delhi, 2015.

Premjit Sharma: *Agricultural Drainage and Water Quality*, Gene Tech Books, Delhi, 2007.

Premjit Sharma: *Agricultural Drainage and Water Quality*, Gene Tech Books, Delhi, 2007.

Przyrbyla, J.: *Practical Considerations for GIS*. Redlands, California: 2010 GeoDesign Summit, 06-08 January 2010.

Punmia, B.C. : *Wastewater Engineering*Laxmi Publications, Delhi, 1998.

Raju, R. Dhana : *Handbook of Mineral Exploration and*

Ore Petrology : Techniques and Applications, The Geological Society of India, Delhi, 2009.

Ramasamy, S.: *Remote Sensing in Water Resources,* Rawat, 2005.

Rao K. Nageswara : *Water Resources Management : Realities and Challenges*, New Century Publication, Delhi, 2006.

Ritter, D.F., Kochel, R.C., and Miller, J.R.: *Process Geomorphology*, Wm.C. Brown Publishers, Dubuque, Iowa, 2005.

Robb, L.: *Introduction to Ore-Forming Processes,* Blackwell Science, Delhi, 2005

Robert, P.: *Organic Metamorphism and Geothermal History*, Elf-Aquitaine and D. Reidel Publishing Company, Boston, 2008.

Robinson, I. S.: *Satellite Oceanography: An Introduction for Oceanographers and Remote-sensing Scientists*, New York, Halsted Press, 2005.

Robinson, I. S.: *Satellite Oceanography: An Introduction for Oceanographers and Remote-sensing Scientists*, New York, Halsted Press, 1985.

Rucker, Rudolph: *Geometry, Relativity and the Fourth Dimension,* NY: Dover, 1977.

Sachanen, A. N.: *Conversion of Petroleum: Production of Motor Fuels by Thermal and Catalytic Processes*, Reinhold Pub. Corp., New York.

Samuel, G. Robello: *Downhole Drilling Tools: Theory and Practice for Engineers and Students,* Gulf Pub., Houston, 2017.

Sanghvi Sheela and Pahwa Prem S. : *Water Harvesting, Purification and Distribution Management* , Dominant, Delhi, 2001.

Scheidegger, Adrian E.: *Morphotectonics*, Springer, Berlin, 2004.

Scheidegger, Adrian E.: *Morphotectonics*, Springer, Berlin, 2014.

Scott, A.C.: *Coal and Coal-bearing Strata: Recent Advances*, Blackwell Scientific Publications, Geological Society Special Publication, Boston, 2007.

Sedimentary, and Metamorphic, New York: W. H. Freeman, 2005.

Seidel, A.: *Kirk-Othmer Encyclopedia of Chemical Technology*, Wiley-Interscience, New York, 2014.

Selby, Michael John: *Earth's Changing Surface: an Introduction to Geomorphology*, Clarendon Press, Oxford, 2005.

Selby, Michael John: *Earth's Changing Surface: an Introduction to Geomorphology*, Clarendon Press, Oxford, 1985.

Shamsi, U.M.: *GIS Applications for Water, Wastewater, and Stormwater Systems*, Routledge, Delhi, 2010.

Sharma, V K : *Origin and Development of Geology*, Vista International Pub, Delhi, 2008.

Sharma, V K: *Origin and Development of Geology*, Vista International Pub, Delhi, 2016.

Singh Chandra Prakash: *Applied Geomorphology : A Study*, B R Pub, Delhi, 2002.

Singh, A.K.: *Environment and Water Resources Management*, Adhyayan, Delhi, 2006.

Smithson, Peter: *Fundamentals of the Physical Environment*, Routledge, London, 2012.

Sofer, Arnon, *Rivers of Fire: the Conflict over Water in the Middle East*, Rowan & Littlefield Publishers, Lanham, MD, 1999.

Spitz, Peter H.: *Petrochemicals: The Rise Of An Industry,* Wiley InterScience, Hoboken, N.J., 2008.

Stach, E., M.-Th. Mackowsky, M. Teichmuller, G.H. Taylor, D. Chandra, and R. Teichmuller: *Stach's Textbook of Coal Petrology*, Gebruder Borntraeger, Berlin & Stuttgart, 2008.

Stanley, Steven M.: *Earth System History*, W.H. Freeman, UK, 1999.

Strahler, Alan; Strahler Arthur: *Introducing Physical Geography*, Wiley, New York, 2006.

Suarez-Ruiz, I., and J.C. Crelling: *Applied Coal Petrology: the Role of Petrology in Coal Utilization*, Academic Press, New York, 2016.

Summerfield, Mike: *Global Geomorphology*, Longman, London, 1991.

Summerfield, Mike: *Global Geomorphology*, Longman, London, 2011.

Swarnim, K. : *A Textbook of Biochemistry and Microbiology,* Surendra Pub, Delhi, 2010.

Thomas, Harold E.: *The Conservation of Groundwater: A Survey of the Present Groundwater Situation in the U.S.,* McGraw Hill, New York, NY. 1951.

Thomas, L.: *Handbook of Practical Coal Geology*: John Wiley & Sons, Chichester, 2012.

Thompson, Graham R. and Turk, Jonathan: *Modern Physical Geology*, Saunders College Publishing, 1991.

Tideman, E. M.: *Watershed Management – Guidelines for Indian Conditions*, Omega Scientific Publishers, Delhi, 1996.

Tissot, B., and D.H. Welte: *Petroleum Formation and Occurrence*, Springer-Verlag, New York, 2014.

Todd, David Keith: *Groundwater Hydrology*. John Wiley & Sons, New York, NY. 1959.

Trivedy, R.K. and Siddharth Kaul: *Low Cost Wastewater Treatment Technologies*, ABD Publishers, Delhi, 2001.

Turcotte, D.L.; Schubert, G.: "Plate Tectonics" *Geodynamics,* Cambridge University Press, Cambridge, 2002.

Tyson, R.V.: *Sedimentary Organic Matter; Organic Facies and Palynofacies*, Chapman & Hall, New York, 2015.

Viles, Heather: *Biogeomorphology*, Basil Blackwell, Oxford, 1988.

Wainwright, John; Mulligan, M.: *Environmental Modelling: Finding Simplicity in Complexity*, John Wiley and Sons Ltd, London, 2013.

Wainwright, John; Mulligan, M.: *Environmental Modelling: Finding Simplicity in Complexity*, John Wiley and Sons Ltd, London, 2003.

Walas, S. M.: *Phase Equilibria in Chemical Engineering,* Butterworths, Reading, MA, 2015.

Watson, O. Michael : *Symbolic and Expressive use of Space: An Introduction to Proxemic Behavior,* New York, *1972.*

William, K. : *Irrigation Investment Technology, and Management Strategies for Development Studies in Water Policy Management 9*, Boulder, Colo Westview Press, 1986.

Winchester, Simon: *Krakatoa: The Day the World Expolded - August 27 1883,* HarperCollins, New York, 2004.

Wolf, P. R.: *Elements of Photogrammetry,* New York, 1983.

Wolf, P. R.: *Elements of Photogrammetry,* New York, 2013.

Yang, Y., R. Zou, Z. Shi, and R. Jiang: *Atlas for Coal Petrography of China*, China University of Mining and Technology Press, Beijing, 2006.

Zim, Herbert S. and Paul R. Shaffer: *Rocks and Minerals: A Guide to Familiar Mnierals, Gems, Ores and Rocks,* Golden, New York, 2007.

Ziminska, H.: *Protein Recovery from Fish Wastewaters*; American Society of Agricultural Engineering; MI, U.S.A., 1985.

INDEX

C

D

F

G

V

W

Z